W0234689

Learning to Live with Climate Change

This imaginative and empowering book explores the ways that our emotions entangle us with climate change and offers strategies for engaging with climate anxiety that can contribute to social transformation.

Climate educator Blanche Verlie draws on feminist, more-than-human and affect theories to argue that people in high-carbon societies need to learn to 'live-with' climate change: to appreciate that human lives are interconnected with the climate, and to cultivate the emotional capacities needed to respond to the climate crisis. *Learning to Live with Climate Change* explores the cultural, interpersonal and sociological dimensions of ecological distress. The book engages with Australia's 2019/2020 'Black Summer' of bushfires and smoke, undergraduate students' experiences of climate change, and contemporary activist movements such as the youth strikes for climate. Verlie outlines how we can collectively attune to, live with, and respond to the unsettling realities of climate collapse while counteracting domineering ideals of 'climate control.'

This impressive and timely work is both deeply philosophical and immediately practical. Its accessible style and real-world relevance ensure it will be valued by those researching, studying and working in diverse fields such as sustainability education, climate communication, human geography, cultural studies, environmental sociology and eco-psychology, as well as the broader public.

Blanche Verlie is an Australian climate change educator and researcher currently living on unceded Gadigal Country. Blanche has over 10 years' experience teaching sustainability and climate change in universities, as well as experience in community-based climate change communication and activism. She has a multidisciplinary background, brings an intersectional feminist approach to her work and is passionate about supporting people to engage with the emotional intensities of climate change. Blanche is currently completing a Postdoctoral Fellowship at the Sydney Environment Institute at the University of Sydney.

Routledge Focus on Environment and Sustainability

Post-Pandemic Sustainable Tourism Management
The New Reality of Managing Ethical and Responsible Tourism
Tony O'Rourke and Marko Koščak

Consumption Corridors
Living a Good Life within Sustainable Limits
Doris Fuchs, Marlyne Sahakian, Tobias Gumbert, Antonietta Di Giulio, Michael Maniates, Sylvia Lorek and Antonia Graf

The Ecological Constitution
Reframing Environmental Law
Lynda Collins

Effective Forms of Environmental Diplomacy
Leila Nicolas and Elie Kallab

Coastal Wetlands Restoration
Public Perception and Community Development
Edited by Hiromi Yamashita

Sustainability in High-Excellence Italian Food and Wine
Laura Onofri

Learning to Live with Climate Change
From Anxiety to Transformation
Blanche Verlie

For more information about this series, please visit: https://www.routledge.com/Routledge-Focus-on-Environment-and-Sustainability/book-series/RFES

Learning to Live with Climate Change

From Anxiety to Transformation

Blanche Verlie

Routledge
Taylor & Francis Group
LONDON AND NEW YORK

earthscan
from Routledge

First published 2022
by Routledge
2 Park Square, Milton Park, Abingdon, Oxon OX14 4RN

and by Routledge
605 Third Avenue, New York, NY 10158

Routledge is an imprint of the Taylor & Francis Group, an informa business

© 2022 Blanche Verlie

British Library Cataloguing-in-Publication Data
A catalogue record for this book is available from the British Library

Library of Congress Cataloging-in-Publication Data
Names: Verlie, Blanche, author.
Title: Learning to live with climate change : from anxiety to transformation / Blanche Verlie.
Description: Milton Park, Abingdon, Oxon ; New York, NY : Routledge, 2022. | Series: Routledge focus on environment and sustainability | Includes bibliographical references and index.
Identifiers: LCCN 2021013039 (print) | LCCN 2021013040 (ebook) | ISBN 9780367441258 (hardback) | ISBN 9781032073668 (paperback) | ISBN 9780367441265 (ebook)
Subjects: LCSH: Climate change. | Human beings--Effect of climate on. | Communication in social action.
Classification: LCC QC903 .V46 2022 (print) | LCC QC903 (ebook) | DDC 304.2/5--dc23
LC record available at https://lccn.loc.gov/2021013039
LC ebook record available at https://lccn.loc.gov/2021013040

ISBN: 978-0-367-44125-8 (hbk)
ISBN: 978-1-032-07366-8 (pbk)
ISBN: 978-0-367-44126-5 (ebk)

DOI: 10.4324/9780367441265

Typeset in Times New Roman
by MPS Limited, Dehradun

Contents

Acknowledgements   vi

1   Introduction: Climate is living-with   1

2   Feeling the climate crisis   22

3   Encountering climate anxiety   46

4   Witnessing multiple climate realities   65

5   Storying climate collectives   88

6   Conclusion: Bearing worlds   111

Appendix: Discussion questions   126
Index   128

Acknowledgements

The research for and writing of this book was written on stolen and unceded Wurundjeri, Boon Wurrung, Dja Dja Wurrung and Gadigal Country. I pay respects to all Australian Aboriginal and Torres Strait Islander elders, past, present and yet to come, for their continued leadership caring for their communities, culture and Country. These places that I have lived in and with are vital contributors to this book.

This book stems from the research conducted for my PhD at Monash University. My supervisors, Iris Duhn and Lauren Rickards (from RMIT University), provided impassioned, provoking, challenging and encouraging discussion throughout the four years of my candidature. Without their tireless work, this book would never have even been imagined. Lauren was also the course coordinator and lecturer in the classes this book discusses, and her guidance in this role was uplifting.

My students from RMIT University are the motivation for this book. A special thanks to those from Climate Change Responses in 2015, whose stories, experiences and voices are the centrepiece of this book. Collectively, your courage and curiosity for learning with climate change inspired me and continues to do so. Thank you all for your willingness to openly discuss your experiences with me.

My PhD examiners, Astrida Neimanis and Marcia McKenzie, provided careful critical analysis of that thesis which has informed the development of this manuscript. Thank you for the time and care taken to read and respond to my work.

Tim Shue from ICLEI and Chanel Koeleman from RMIT University helped me develop ideas from the PhD into a concise article for *The Conversation* (https://theconversation.com/the-terror-of-climate-change-is-transforming-young-peoples-identity-113355)

After reading that article, Annabelle Harris from Routledge invited me to submit a proposal for this book. I would not have had the self-belief to do so otherwise; thank you for seeing the potential of my work and encouraging me on this path. Matthew Shobbrook patiently guided me through the publication process, with comprehensive and prompt responses even in the midst of the devastating COVID-19 pandemic.

This manuscript has been written entirely during the pandemic, while I was completing my postdoctoral fellowship at the Sydney Environment Institute at the University of Sydney. Thank you to David Schlosberg, Astrida Neimanis, Dany Celermajer, Dinesh Wadiwel and Sophie Webber for the academic support during this testing time.

This book argues that our capacities emerge from our relationships; my abilities are no exception. Throughout the last few years, I have been fortunate to be supported by a vast range of people, networks, and institutions. In addition to those already mentioned and in no particular order: the Sydney Environment Institute (especially Omar Elkharouf, Eva Perroni, Anna Sturman, Chris Dundas, Stella Maynard and Catriona Macmillan); the Sustainability and Urban Planning and Centre for Urban Research staff at RMIT University (especially Libby Porter, Susie Moloney, Wendy Steele, Karyn Bosomworth, Briony Towers, Bronwyn Lay, Ben Cooke, Melissa Neave, Mittul Vahanvati, Nia Emmanouil, Steven Jeffrey, David Meiklejohn and those who inspired my love of teaching in the first place: Anne-Lise Ah-Fat, Arley Marks, Laurel Mackenzie and Gabriella Corbo-Perkins); Composting Feminisms (convened by Astrida Neimanis and Jen Hamilton); Eco-Feminist Fridays (convened by Hayley Singer); the Hacking the Anthropocene IV organising collective (Hayley Singer, Steph Lavau, Anna Dunn and Tessa Laird); editors of special issues and anonymous peer reviewers of articles I have published; the New Ecological Discourses reading group (convened by Freya Mathews); #aaeeer (especially Claudio Aguayo, Maia Osborne, Sherridan Emery, Kim Beasy and Kevin Kezabu); other academics and PhD students at Monash University (especially Sarita Galvez, Misol Kim, Mark Boulet, Yoshi Nakagawa, Sue Grieshaber and Alan Reid); Climate for Change (especially Oli Moraes, Zey Basarin, Anna Dunn, Nathan Eizenberg, Gabriella Maher, Nicole Robertson, Jimmy Ingram, Ajaya Haikerwal and Katerina Gaita); Climates (especially Jarrod Troutbeck, Jeremy Gay, Malcolm Roberts and Tomas Sanchez); Psychology for a Safe Climate (especially Carol Ride and Bronwyn Wauchope); my co-conspirator for FireFeels, Mandy Pritchard; co-authors and co-editors Alicia Flynn, Simone Blom, Tamara Jarrett, Emily Clark and Emma Supriyono; the *Australian Journal of Environmental Education* Editorial Executive (Amy Cutter-MacKenzie-Knowles, Peta White, Chris Eames and Marianne Logan); the Climate Change and Rural Mental Health team at the University of Sydney (Jo Longman and Maddy Braddon); the Multispecies Justice research community (especially Christine Winter and

Sophie Chao); and the Climate Change and Educational Institutions research team (Jeff Brooks and Lisa Kensler).

Dear friends have helped me think harder, relax, have fun, get out of the city, learn more and keep at it across these last few years. Thank you Jessie Hui, Jace Loh, Gary Littlejohns, Tom Backay, Helena Puszka, Val Romanov, Lorraine Dawes, Anna Romanov, Geordie Williamson, Bridget Malcolm, Susie Pratt, Tema Milstein, Laura McLaughlan and Lindsay Kelley.

David Schlosberg, Dany Celermajer, Mandy Pritchard, Anna Dunn, Astrida Neimanis, Oli Moraes, Meddwyn Coleman, Hayley Singer and Nathan Eizenberg provided helpful written comments on the book manuscript. Colleagues from the Sydney Environment Institute and Urban Crew at the University of Sydney provided important feedback on early plans for the book.

My family, Trevor, Morgan, Katie and Kai Coleman, and especially my mum, Meddwyn Coleman, and partner, Rob Goodwin, and most especially, my canine companions, Bella and Jessie, have offered unwavering support of all the best and most important kinds: food, hugs, walks outside and home. You're the best.

Some small sections of this book have been reproduced from the following articles and are reproduced with permission from the publishers: Taylor and Francis (papers 1–3 below), Elsevier (4) and Cambridge University Press (5).

Verlie, Blanche. "Rethinking Climate Education: Climate as Entanglement." *Educational Studies* 53, no. 6 (2017): 560–72. https://doi.org/10.1080/00131946. 2017.1357555.

Verlie, Blanche. "Bearing Worlds: Learning to Live-with Climate Change." *Environmental Education Research* 25, no. 5 (2019): 751–66. https://doi.org/ 10.1080/13504622.2019.1637823.

Verlie, Blanche, and CCR15. "From Action to Intra-Action? Agency, Identity and 'Goals' in a Relational Approach to Climate Change Education." *Environmental Education Research* 26, no. 9–10 (2020): 1266–80. https://doi. org/10.1080/13504622.2018.1497147.

Verlie, Blanche. "'Climatic-Affective Atmospheres': A Conceptual Tool for Affective Scholarship in a Changing Climate." *Emotion, Space and Society* 33 (2019): 100623. https://doi.org/10.1016/j.emospa.2019.100623.

Verlie, Blanche, Emily Clark, Tamara Jarrett, and Emma Supriyono. "Educators' Experiences and Strategies for Responding to Ecological Distress." *Australian Journal of Environmental Education* Online First (2020): 1–15. https://doi.org/ 10.1017/aee.2020.34.

1 Introduction: Climate is living-with

'I'm fuming' a colleague tells me. She is speaking about the cata-strophic fires raging across Australia, which have sent massive plumes of smoke billowing into our skies, with only a pitiful response from our government.

'When we've got no water in our rivers, it feels like we're drained as well' says Elder Vanessa Hickey in a report about chronic drought leaving towns without a drop.[1]

Timothy Morton refers to Hurricane Harvey 'unleashing Hurricane Tim,' as the atmospheric chaos reverberates throughout their household.[2]

A marine biologist vomits because of her distress about coral bleaching, mimicking her beloved polyps who purge themselves of their symbiotic algae in warming water.[3]

Rebecca Huntley recounts a sensation that 'actually felt *physical*, as if vital organs had moved inside my body' when watching youth climate activists implore adults to 'do something.'[4]

School strikers declare: 'as oceans rise, so do we!'

***

The more I search, the more examples I find of such affective entanglements between climates and their humans, even before we consider climate change. The Anglophone world has led humankind in fossil fuel extraction and climate denial. Yet even the English language abounds with idioms that express sentiment through meteorological metaphors. We might feel under the weather, right as rain, or on cloud nine. Atmospheric figurations also express broader dispositions such as having your head in the clouds, making hay while the sun shines, or throwing caution to the wind, as well as interpersonal relations like stealing someone's thunder or being fair weather friends. Temperature,

DOI: 10.4324/9780367441265-1

Tim Ingold reminds us, has the same linguistic root as temper.[5] We often use 'atmosphere' and 'climate' to refer to the aura, vibe or affective patterns of a place, time, group or event. These linguistic practices seem to suggest that we understand that human experiences are intimately related to the meteorological. In places like Melbourne, Australia, where I lived for over a decade, we start almost every conversation with the weather, because we implicitly know it affects each other's mood and wellbeing. We have all had experiences when the weather gave us a spring in our step, left us feeling exhausted or made us want to go back to bed.

Yet our capacity to feel is rarely acknowledged as a legitimate way of knowing climate change. Public and academic approaches to human-climate relations still tend to normalise and advocate scientific modes of climate knowledge, which promote mental comprehension of statistics and graphs through disembodied abstraction. In Australia and other places where climate action has been stymied by systemic and institutionalised denial, activists (of all sorts) have worked to firm up the borders and reputation of science, and distinguish it from ideological, irrational, and emotional 'post-truth' regimes. However, positioning climate change as a phenomenon to be known primarily through science has led to approaches to public engagement that are highly disengaging, as well as ignoring the emotional pain of those who are already concerned.

Research is increasingly finding that climate denial's apparent opposite, climate anxiety, is one of the major barriers to climate action. Indeed, what appears to be apathy can actually be feelings of grief and disempowerment that are too difficult to engage with, leading to denial as a mechanism for short-term emotional coping.[6] If there is a lack of care, it is not that most of us do not care, but that we do not know how to care. We do not have the inter/personal competencies necessary for engaging with the intense combination of guilt and fear induced by this existential crisis.

I believe that if we are to adequately respond to climate change, we need to consider humans' ability to feel climate as a serious and powerful mode of engagement. Beginning from this premise, in this book I develop a pedagogy (a theory of learning, and strategies for fostering it) that centres human feelings as potent apparatuses for knowing climate. This approach can better attune to the intimate ways people are enmeshed with climate and cultivate the emotional capacities required for facing climate collapse.

However, before we can explore this, we need to begin by reconsidering what we understand climate to be, because this underpins our pedagogies. Our dominant understandings of *what climate is* are derived from Western climate science, and although they offer us absolutely critical knowledge, they fail to adequately engage or support people in the transformations that climate science itself tells us are necessary. Western science understands climate as a system composed of interacting subsystems: the biosphere, atmosphere, hydrosphere, cryosphere (frozen water) and the lithosphere (rocks and soil). Climate is therefore also a statistical measure of these interactions which is calculated through the average conditions of the weather – such as the temperature, humidity and pressure of the atmosphere – and is usually measured over periods of 30 years. Climate change is thus a change in the average conditions of the atmosphere, and this is taken to represent changes in the whole climate system.

These scientific understandings have failed to engage the masses in part because they do not offer relatable, connective or inspiring accounts of human-climate relationships. To be fair, climate science is increasingly using 'Gaia-like principles' of 'processes and relations'[7] to represent the climate, and humans are incorporated, as a homogeneous species, into the models scientists use to create projections about potential future climatic changes. However, while climate science seeks to inform us of humanity's connection with the climate, the climate scientists themselves are considered to be external observers of the climate system, occupying a 'God's eye view' of the planet.[8] The mechanistic language of systems combined with the myth of the invisible scientist continue to perpetuate assumptions that living, flesh-and-blood individual humans are separate from the climate. We are offered two depictions of the human within human-climate relations: the disembodied and dispassionate scientist, or the faceless masses of 'humanity' represented as numbers within computer models. Most people do not relate to either. Neither offers a template for living in and with climate as a real-life embodied person; nor do they account for the diverse possibilities of ways of being human.

Thankfully, climate change engagement has moved beyond these dissociative accounts. Alternative approaches have largely been informed by social constructionism, which believes that people create knowledge and meaning through social interaction and cultural engagement. Such approaches allow us to investigate and appreciate the influential roles of ideology and culture in how we understand and thus relate to climate. They have greatly improved strategies for engaging people with climate change. Indicative of these approaches, Mike Hulme, a world-leading climate change scholar, argues that scientific definitions of climate:

too easily maintain a false separation between a physical world (to be understood through scientific inquiry) and an imaginative one (to be understood through meaningful narratives or human practices). Such a distinction maps easily onto the nature-culture dualism which has engrained itself in much western thought and practice.[9]

In response, Hulme has contended that climate 'needs to be understood, first and foremost, culturally'[10] and as an 'idea of the human mind.'[11] However, this does not overcome the human/nature dualism that Hulme rightly critiques. Rather, it switches from one side to the other, enclosing climate entirely within the human imagination.

As a student of mine once stated, 'I don't believe climate can be a purely constructed idea because we can feel it, not just with our senses but with our emotions and in our subconscious.' The Bawaka Country research collective add that 'the patterns of climate exist far beyond human minds, have their own existence and agency, have knowledge and law, even as they are entangled with us.'[12] While both scientific and social constructionist approaches have value, neither fully engages with the human body as a legitimate means of knowing climate change.[13] Similarly to scientific approaches, social constructionism erases human embodiment and considers humans to be onlookers or inventors of climate rather than intimately enmeshed within its living material fluxes.[14] As Indigenous and posthuman[15] scholarship argues, this is because both Western science and social constructionism are founded on problematic anthropocentric, dualistic and individualistic assumptions.[16] Anthropocentrism is the belief that humans are separate or separable from, unique in comparison to, and more important than, the non-human world. Some of our apparent unique qualities include the possession of minds (and thus consciousness, spirit, knowledge and intelligence) and of agency (the ability to influence change in the world). The supposed exceptionalism of the human mind leads to a denial of the value and capacities of the intelligence of human bodies. This myth of human cognitive supremacy also works to justify the belief that humans can and should fully comprehend and control the non-human world.

Unfortunately, while such anthropocentrism is acknowledged as one of the root causes of ecological crises,[17] it remains the philosophical foundation for almost all climate change engagement efforts (i.e. education, communication, awareness building, campaigning, behaviour change and related policy).[18] In our aspirations to reduce emissions, we perversely perpetuate deep seated extractivist understandings of humans as autonomous, entrepreneurial selves who, if only they cared enough, would be able to exert morally righteous control over themselves, their actions, and therefore, the non-human world. This limits our abilities to understand ourselves as part of climate, to engage with our embodied experiences of climate change, and to cultivate collective climate action. But what if we began with a different understanding of *what climate is*, and how we can know it? What if lived, embodied, emotional, inter-personal and relational experiences were considered *constitutive of* climate and as *valuable ways to comprehend it*?

This book argues that we need to learn to live with climate change. Learning to live with climate change begins from an understanding that *climate is living-with*. This is climate not as a noun referring to a thing – whether a cultural idea or a geophysical system – but climate as a verb, referring to an action: to processes of living-with ourselves, others, and the world. Climate is something we all do, all the time, and we are always doing it together, even if we are not doing it in coordinated, similar or equal ways. Climate as living-with focuses on the intimate ways we are entangled with the non-human world, and how the patterns of these relationships generate the conditions in which we live. It therefore attunes to how the planetary and epochal phenomenon of climate change is metabolically, emotionally and politically enmeshed within our everyday, mundane, inter/personal lives and compels respect, reciprocity and responsibility for this expansive relationality. This understanding better enables us to culti-vate strategies that can adequately inspire and support people to engage with and respond to climate change, and to do this in ways that contribute to multispecies climate justice.[19]

In figuring climate as living-with, I am drawing upon critical feminist posthumanism,[20] cultural geography's articulation of the affectivity of atmospheres[21] and, to the extent that I am able to do so, Indigenous philosophies.[22] More specifically, I am indebted to Astrida Neimanis' and colleagues' notion of 'weathering' that poetically articulates the ways in which our different bodies are enmeshed – differently – with climate,[23] and to Nancy Tuana who demonstrates that lived experiences of disasters can help us cultivate awareness of this.[24] I am fascinated by Waanyi writer Alexis Wright's exploration of the awesome agency of the atmosphere and enlivened by her suggestion that even settler people have some remnant appreciation of this.[25] Similarly, I am captivated by Métis scholar Zoe Todd's articulation of climate as a 'sentient commons,'[26] although I am aware that much of what this nuanced phrase offers is probably lost on me. Building on these relational understandings of climate, Potawatomi scholar Kyle Powys

Whyte's exploration of the qualities of relationships (such as consent, trust, accountability, and reciprocity) that constitute livable climates and which must be re-established for climate justice is central to the ethos of living-with climate.[27] I am buoyed by Ashlee Cunsolo's elucidation of the transformative politics of mourning climate change.[28]

Common to such approaches, and thus to this book's philosophy of climate as living-with, are four key interdependent notions. The first is about interconnection, entanglement or *relationality*: that to exist is to be composed, and continually re-composed, through relationships with others, and that climate is not an object so much as a patterned 'set of relations.'[29] Second, these relationships are always *more-than-human*: we cannot escape our entanglement with climate and the wider ecological world. Relatedly, it is not just humans that change climate; non-humans also participate in creating, stabilising and changing climate, although this does not discount the significance of the changes being wrought by some human systems in this geological moment. Third, climate is *embodied*, and all earthly beings are viscerally enmeshed with climate; indeed, we become (with) climate.[30] Fourth, climatic phenomena are inherently *affective*: they are energies, forces, intensities, feelings. Collectively, these principles articulate climate as 'a living phenomenon'[31] that emerges from the interactions and relationships between all bodies: human, non-human and 'inanimate'; living, dead, ancient and yet to come. In describing climate as living-with – as patterns of affect; as flows of feeling; as repertoires of re-lating; as a sensational phenomenon; as multispecies enmeshment – I am seeking to expand how we might relate to climate, rather than narrow this. Thus, I take a capacious approach to these explorations because part of my agenda is to enable openness to climate's exces-siveness, especially in ways that I have not anticipated.

Attuning to climate as a process of living-with tries to get out of the damaging 'worldview' that literally gazes upon the world as though it were an object external to the self and available for human use. Climate as living-with conceptualizes humans and atmospheres as temporary differentiations of the same visceral materiality which cycles through the world and which contributes to us all changing through and because of our interpermeation. As a cycle of energetic dispersal, climate affects bodies, is archived within them[32] and is composed by them. Climate is thus not only the conditions for life, but a product of it, and it emerges from all earthly bodies living-with each other. This of course also in-volves patterns of dying; living-with is always also about living-off, living-through and living-after others, as well as knowing that others benefit from our own edibility and mortality. Death, destruction and decomposition are necessary and life-giving parts of what constitutes climate.[33] However, contemporary climate change is a radical disruption in our patterns of relations, leading to what Danielle Celermajer terms *omnicide*: the killing of everything.[34] Omnicidal cultures stem from a failure to recognise our implication within and vulnerability to the more-than-human. For those of us who are complicit with climate change, these are the cultures we have inherited and which we are implicated in.

Of course, this 'we' is not everyone. The 'we' that I am most often referring to in this book is, except where specified otherwise, those of us who are significantly complicit with climate change. To be climate-complicit is to benefit from climate-changing systems: capitalism, colonialism, industrialisation and/or extractivism more broadly. Being a beneficiary of such systems does not preclude being distressed and threatened by climate change. Indeed, climate change can leave us feeling deeply *unsettled* because it disrupts the sense of security – comfort, control, complacency – that global histories of imperialism have afforded us. Although not all climate-complicit people are living on stolen Indigenous land, at times I therefore refer to this 'we' as settlers. In doing so I am using this term in an expansive way in order to attune to the trans-national affective regimes of extractivism that contribute to climate-complicit people's eco-anxiety.[35] This 'we' is an admittedly ambiguous collective, and I do not try to draw tight parameters around it other than clarify that I am not referring to all of humanity when I argue that if 'we' are to become climate-responsible, we will need to *learn* to live with climate change.[36]

We, those of us steeped in extractivist cultures, urgently need to cultivate a relational understanding of climate, which is to say, we need to consider humans as part of climate, and to appreciate that life is always living-with climate. While this can sound like a romantic or rose-tinted approach, in an era of climate crisis and ecological collapse, being interconnected with nature is not a choice, nor is it inherently nice: it is our interrelatedness that makes our organs fail in extreme heat, leaves local economies reeling from cyclones, and leads to complex intergenerational grief when ancestral homelands are slowly eaten away by the rising tides. And it is through our inter-connectedness with climate that we are making this happen.

Appreciating our intimate relationality with climate change is therefore deeply distressing. Understanding climate as living-with acknowledges we can and do feel violences inflicted on the atmosphere and broader planetary relations in our own bodies, as these violences are also inflicted, in some ways, on ourselves. For example, in the undergraduate class on climate change I taught for five years, in one semester my students stated that climate change made them feel anxious, frustrated, confused, uncertain, cynical, scared, overwhelmed, emotional, devastated, depressed, frightened, angry, gloomy, resentful, challenged, isolated, desperate, disheartened, shocked, concerned, confronted, unsettled, bitter, sad, sick, upset, perplexed, guilty, stressed, amazed, daunted, defeated, dismayed, pessimistic, uneasy, tired, appalled and terrified. Given the incomprehensibly rapid and traumatic changes being wrought upon our planet's climate, it is unsurprising that many of us are overwhelmed with climate anxiety, whether we can consciously and reflectively acknowledge this or not. Kari Norgaard argues that if we acknowledge that climate change is *'too disturbing* to be fully absorbed or integrated into daily life' then denial and apathy can 'be understood as testament to our human capacity for empathy, compassion, and an underlying sense of moral imperative to respond, even as we fail to do so.'[37] Thus, we need to be able to muster the courage to face up to our vulnerability and complicity in climate change, painful as it is, because it is only from there that we will be able to transform ourselves and our worlds. We must cultivate an ethos of *living-with* – respecting, being part of, enduring and responding to – climate change.

Research investigating the mental health impacts of climate change and how to build emotional resilience to this is increasing.[38] However, I worry that a focus on emotional resilience risks re-centring the individual human – normally privileged ones – and breeding acceptance in place of outraged inspiration. In contrast, I want to rethink how we understand 'the human' and how humans are, and can be differently, entangled with/in climate change. I want an approach that is less individualistic, less anthropocentric, and more ecological, more collective. I want responses to climate change that are both humbler and more ambitious. I want an approach that does not help privileged people feel okay or help them accept climate change, but one which enables us to bear the burden of complicity in ways that hold us accountable and that generate radical change.

Rather than emotional resilience, learning to live with climate change aspires for affective transformation. The climate crisis is traumatic because it renders apparent the grotesque manifestations of our unchecked individualistic sense of self. Rather than cultivate tolerance of the unconscionable violences that are being wrought on species, ecosystems, human people and communities, we need to *transform* ourselves and our affective norms and repertoires. We literally need to engage in other forms of being, and ones that do not draw self-enclosing boundaries around the individual human but consider the 'self' to be dispersed in-between and across, and constantly emerging with ('trans'), its relations with others. Affective transformation could enable us to more thoroughly empathise with others near, far, estranged and yet to come, as well as to draw strength and joy from our relations, empowering us to face up to, address and prevent the injustices that climate change engenders. Advocating for affective transformation as a response to complicit people's ecological distress is an effort to cultivate emotional climate justice: to work with emotions for climate justice, and to work towards a more just distribution of the emotional impacts of climate change.

***

To further explore these notions of climate as living-with, Chapter 2 attunes to the affectivity of Australia's 'Black Summer' (in 2019/2020) of extreme bushfires and chronic smoke pollution. The unprecedented intensity and scale of these fires was attributable to climate change,[39] which in turn is attributable to decades of climate apathy and inaction. In this context of smothering denial, the fire season ignited fiery controversy, burning rage, smouldering grief, fuming shock and searing abandonment. The physical, emotional and political ways that the 2019/2020 bushfires intruded into people's bodies and worlds demonstrate that atmospheres and climates are both meteorological and affective, and that people and climates co-compose each other.[40] Considering these various modes of feeling climate change and the public's immediate responses to the 2019/2020 bushfires, this chapter gestures towards the need for careful strategies to support people to learn to live with climate change.

Chapters 3, 4 and 5 take up this challenge and explore key practices through which this can be achieved: encountering, witnessing and storying climate change. These practices are qualitatively different to 'knowing about' climate change, whether through scientific or social constructivist approaches, or 'acting on' it. They are embodied, affective and relational practices that we enact with, and as part of, climate change, and through which we become climate-changed. These three practices are inter-related, and therefore each is threaded through all of the chapters. However, Chapter 3 focuses on how encountering climate anxiety can unsettle the affective norms that sustain extractive cultures. Chapter 4 highlights how witnessing multiple climate realities can enroll us in additional, and alternative, relations with climate change. Chapter 5 attunes to how carefully storying climate collectives can enact more promising futures. Collectively, these practices can generate affective, and thus, social, economic and ecological, transformation. While numerous researchers have argued for the urgent political need to acknowledge, explore and make experiences of ecological distress public,[41] these chapters develop a pedagogy for doing so. Increased awareness of when, where, how and with whom we can and do encounter, witness and story climate change can enable more attuned and responsive climate change engagement strategies.

My exploration of these practices is informed by a decade of experience working as a climate change educator and facilitator with students and communities, as well as personal experiences with friends, family and colleagues, and my participation in and observation of climate activist movements such as Extinction Rebellion, the youth strikes for climate, and the climate emergency mobilisation. These three chapters also include a detailed discussion of one of my university classes, which enables close attention to how these practices overlap and interrelate. The class I focus on was an undergraduate social science climate change class in which I was a tutor with approximately 45 students across two tutorials. In the class we studied the political, economic, social, psychological and technological dimensions of the causes and consequences of climate change, all with a focus on climate in/justice: how those who contribute least to climate change tend to experience the worst impacts of it, and how addressing this injustice requires participatory responses led by those on the front lines. Through focusing on this one class, these three chapters together explore how we can encounter, witness and story the ways we, and others, are encountering, witnessing and storying climate change. For example, through our class discussions, we witnessed each other's encounters with climate anxiety; we subsequently told stories about collectively witnessing each other; and these stories in turn countered our individual and collective sense of self. These spiralling and overlapping practices of engaging with and sharing our climatic experiences contributed to an unsettling, but also regenerative, atmosphere that enabled us to continue in this work.

Encountering, witnessing and storying climate change are not only a pedagogy I explain and advocate, but also the methodology I enact. Across the book, I describe, but also engage in, these practices, including through the vignettes that open each of Chapters 3–5. These three vignettes are my effort to witness experiences of ecological distress which are too often glossed over. Collectively, they tell a story of our class, one which seeks to narrate how affective transformation can cultivate alternative identifications and selves, decomposing neoliberal understandings of human subjects as autonomous and self-contained towards more dispersed, fluctuating, more-than-human gatherings. The vignettes are written in first person but assembled from different people's comments, reflections and experiences from this class in an effort to convey the 'cloudy collective' that emerged throughout that semester (see Chapter 5 for more discussion of this). In that sense, these stories are 'made, [but] not made up'[42] and demonstrate my active effort at storying climate collectives.[43] The vignettes perform a semi-coherent collective 'I,' a fickle narrator whose subjectivity oscillates according to wavering affective entanglements with others and with climate change. While just one example of the kinds of non-anthropocentric characters we need, the vignettes demonstrate that imagining, crafting and enacting more promising ways of being human are possible and that this can be cultivated through collectively encountering, witnessing and storying climate change's affective agency.

My efforts to witness and story my students' encounters with climate change in these ways has in turn countered, rather than affirmed, my supposed status as an external and authoritative researcher. Across the years I have spent doing this work, I have been lost, demoralised, exhausted and deflated, but also inspired and energised. These affective experiences leaked from and infiltrated my relationships with my students and other communities, blurring my own sense of individuality. In order to better story this reality, a few of the statements included in the vignettes are 'mine,' although delineating authorship in this way is contrary to the relational disorientation they are trying to narrate. 'My' experiences arose from the close relations I had with others (human and non-human), just as everyone else's did, making it difficult to discern where one person's voice began and another's ended. The vignettes therefore offer 'a non-linear collaborative narrative in which the voices of individual authors' have become 'entangled, and at some points, indistinguishable.'[44]

In addition to describing and performing these three practices, I am also seeking to enrol you, dear reader, in them. It is my hope that through reading, interpreting and responding to the vignettes – in whatever ways you do – you may gain a little bit of lived experience of what encountering, witnessing and storying climate change can do to us. Perhaps you may find yourself similarly unsettled by the atmospheres they enact, and/or drawn into the cloudy collective that shimmers throughout them. What your experience of this may be, I cannot tell, as these stories will intersect with and reverberate throughout your unique affective world. Attuning to how this unfolds for you and how you respond to that will help cultivate understanding of the analysis offered in the book.

It is worth noting that these vignettes include stories of distress; I encourage you to approach them in a mode that cares for yourself and is responsive to your own ability to engage with the pain of climate change at the moment. As this book documents, climate change is deeply traumatic and while I believe we need to avoid the pitfalls of an individualistic approach to emotional resilience, this is not to say that practices of mindful self-care or professional counselling services have no value.[45] They are necessary, indeed critical, just not sufficient, if we are to collectively learn to live with climate change. I hope that describing, enacting and enrolling you in encountering, witnessing and storying climate change enables appreciation of these three practices' multiple sites and cascading forms, and that it cultivates your ability to attentively step into them in your life and guide others to do so too.

Building on the exploration of the possibilities of encountering, witnessing and storying woven between Chapters 3, 4, and 5, Chapter 6 articulates and advocates for affective transformation. Affective transformation recognises that we are atmospheric beings who are always becoming part of, emerging through, acting-with and contributing to the affective flows of climate change.[46] As such, it suggests that our responses to ecological distress need to ensure that we do not try to 'bounce back' to anthropocentric individualism. Rather, we need to change who we are through, and as a means of, responding to the affective pain of climate change. We need to *bear worlds*, where 'worlds' are understood as complex sets of more-than-human relations, dispositions, practices, structures, perceptions and identities. We need to be able to *endure* the pain that business-as-usual worlds are enacting, in order to *generate* more liveable worlds. Learning to live with climate change is therefore not about resignation or giving up. Rather, it is about engaging with and facing up to the horrific realities of climate change and striving to make things otherwise despite knowing that we may not be able to 'save the world.' Indeed, learning to live with climate change acknowledges that 'the' world is not ending, but 'a' world is, and that some worlds need to end in order to allow others room to breathe.[47] Which worlds we nurture matters.

Thankfully, this affective transformation is already underway, even if we do not know, and cannot predict, where it is headed or who we will become. Outrage at suffocating inaction is generating not only protests, but also alternative aspirations, identities and communities.

Figure 1.1 Protestor at the School Strike 4 Climate in Brisbane, Australia, September 20, 2019. The sign speaks to the protestors' affective affinity with their more-than-human kin, as well as the emotional labour of bearing worlds: working with ecological distress to catalyse collective climate action. CC BY 2.0 School Strike 4 Climate.

As the school striker's sign in Figure 1.1 so poetically indicates, climate capable collectives are organising around empathic kinships with the more-than-human world. Crucially, these collectives are emerging through practices of making ecological distress explicit and public, and thus using it as a source of communal motivation and strength.

This book is written with climate change 'educators' in mind: teachers, activists, communicators, young people, parents, researchers, policy makers, community members, artists, politicians – anyone trying to encourage and support people (including themselves) to engage with and respond to climate change. To that end, Chapter 6 concludes with some tangible guidance for those seeking to engage others in the process of learning to live with climate change,[48] and the book closes with an Appendix that offers some example discussion questions. However, this book does not provide quick fixes, because there aren't any. Rather, it seeks to offer a nuanced account of processes and approaches that can help reorient our collective efforts towards more compassionate and transformative responses to climate change. While written from an educator's perspective, I believe this book has much wider relevance, as it speaks to larger social processes that are happening, and can be amplified, in all realms of human life. My ultimate aspiration for this book is to contribute to society-wide cultural changes that might give us a chance of contributing to livable, and even enjoyable, worlds. As a systemic issue that is progressively killing more and more people, species, ecosystems and livelihoods, if we face up to and engage with these issues, climate change could be the teacher we need to help us learn how to live.

Notes

1 Lorena Allam and Carly Earl, "For Centuries the Rivers Sustained Aboriginal Culture. Now They Are Dry, Elders Despair," *The Guardian*, January 22, 2019, https://www.theguardian.com/australia-news/2019/jan/22/murray-darling-river-aboriginal-culture-dry-elders-despair-walgett.
2 Timothy Morton, "The Hurricane in My Backyard," *The Atlantic*, 2018, https://www.theatlantic.com/technology/archive/2018/07/the-hurricane-in-my-backyard/564554/.
3 Susanne Moser, "Getting Real About It: Meeting the Psychological and Social Demands of a World in Distress," in *Environmental Leadership*, ed. Deborah Gallagher (Los Angeles, London, New Delhi, Singapore, Washington DC: SAGE Publications, 2012).
4 Rebecca Huntley, *How to Talk About Climate Change in a Way That Makes a Difference* (Crows Nest and London: Murdoch Books, 2020), 2. (Original emphasis)
5 Tim Ingold, *The Life of Lines* (Milton Park and New York: Routledge, 2015).
6 Kristin Haltinner and Dilshani Sarathchandra, "Climate Change Skepticism as a Psychological Coping Strategy," *Sociology Compass* 12, no. 6 (2018), https://doi.org/10.1111/soc4.12586.
7 Lauren Rickards, "Critiquing, Mining and Engaging Anthropocene Science," *Dialogues in Human Geography* 5, no. 3 (2015): 339, https://doi.org/10.1177/2043820615613263.
8 Donna Haraway, "Situated Knowledges: The Science Question in Feminism and the Privilege of Partial Perspective," *Feminist Studies* 14, no. 3 (1988), https://doi.org/10.2307/3178066.
9 Mike Hulme, "Climate," *Environmental Humanities* 6 (2015): 175–76.

10 Hulme, "Climate," 175.
11 Mike Hulme, *Weathered: Cultures of Climate* (Los Angeles, London, New Delhi, Singapore, Washington, Melbourne: SAGE Publications, 2017), 2.
12 Bawaka Country et al., "Gathering of the Clouds: Attending to Indigenous Understandings of Time and Climate through Songspirals," *Geoforum* 108 (2020): 301, https://doi.org/10.1016/j.geoforum.2019.05.017.
13 Blanche Verlie, "Rethinking Climate Education: Climate as Entanglement," *Educational Studies* 53, no. 6 (2017), https://doi.org/10.1080/00131946.201 7.1357555.
14 Bawaka Country et al., "Gathering of the Clouds: Attending to Indigenous Understandings of Time and Climate through Songspirals."
15 Posthumanism is a school of philosophy that can be quite broad. I draw on that which seeks to challenge anthropocentrism by a) emphasising that humans are dynamically enmeshed with the non-human world and therefore always becoming-with it, and b) deconstructing and resisting hierarchies that position humans as inherently more valuable, intelligent or capable than other beings.
16 Peter Cole, "Education in an Era of Climate Change: Conversing with Ten Thousand Voices," *TCI (Transnational Curriculum Inquiry)* 13, no. 1 (2016); Rosi Braidotti, *The Posthuman* (Cambridge and Malden: Polity Press, 2013); Karen Barad, *Meeting the Universe Halfway: Quantum Physics and the Entanglement of Matter and Meaning* (Durham and London: Duke University Press, 2007).
17 Val Plumwood, *Feminism and the Mastery of Nature* (London: Routledge, 1993).
18 Affrica Taylor, "Beyond Stewardship: Common World Pedagogies for the Anthropocene," *Environmental Education Research* 23, no. 10 (2017), https://doi.org/10.1080/13504622.2017.1325452; Verlie, "Rethinking Climate Education: Climate as Entanglement."
19 Danielle Celermajer et al., "Multispecies Justice: Theories, Challenges, and a Research Agenda for Environmental Politics," *Environmental Politics* 30, no. 1–2 (2020), https://doi.org/10.1080/09644016.2020.1827608.
20 Stacy Alaimo, "Trans-Corporeal Feminisms and the Ethical Space of Nature," in *Material Feminisms*, ed. Stacy Alaimo and Susan Hekman (Bloomington: Indiana University Press, 2008); Barad, *Meeting the Universe Halfway: Quantum Physics and the Entanglement of Matter and Meaning*; Donna Haraway, *When Species Meet* (Minneapolis: University of Minnesota Press, 2008); Donna Haraway, *Staying with the Trouble: Making Kin in the Chthulucene* (Durham and London: Duke University Press, 2016).
21 Ben Anderson, "Affective Atmospheres," *Emotion, Space and Society* 2, no. 2 (2009), https://doi.org/10.1016/j.emospa.2009.08.005; Peter Adey, "Air/Atmospheres of the Megacity," *Theory, Culture & Society* 30, no. 7–8 (2013), https://doi.org/10.1177/0263276413501541; Derek McCormack, "Engineering Affective Atmospheres on the Moving Geographies of the 1897 Andrée Expedition," *Cultural Geographies* 15, no. 4 (2008), https://doi.org/10.1177/1474474008094314; Kathleen Stewart, "Atmospheric Attunements," *Environment and Planning D: Society and Space* 29, no. 3 (2011), https://doi.org/10.1068/d9109.

22 I am a white settler-Australian, and the risks of appropriation, tokenism and misrepresentation loom large in my efforts here. Erasure is the other side of this dilemma. I hope that this book demonstrates that climate change provides another rationale, and a productive opportunity, to work towards decolonisation. I do not pretend to define or map the many processes needed for decolonisation or decolonial climate action; that can only be led by Indigenous peoples. Nor do I pretend to hold Indigenous knowledge but I endeavour to learn from that which is shared with me by Indigenous peoples.

23 Astrida Neimanis and Jennifer Mae Hamilton, "Weathering," *Feminist Review* 118, no. 1 (2018), https://doi.org/10.1057/s41305-018-0097-8; Astrida Neimanis and Rachel Loewen Walker, "Weathering: Climate Change and the "Thick Time" of Transcorporeality," *Hypatia* 29, no. 3 (2014), https://doi.org/10.1111/hypa.12064.

24 Nancy Tuana, "Viscous Porosity: Witnessing Katrina," in *Material Feminisms*, eds. Stacy Alaimo and Susan Hekman (Bloomington: Indiana University Press, 2008).

25 Alexis Wright, "Deep Weather," *Meanjin* 2 (2011), https://meanjin.com.au/essays/deep-weather/.

26 Zoe Todd, "An Indigenous Feminist's Take on the Ontological Turn: 'Ontology' Is Just Another Word for Colonialism," *Journal of Historical Sociology* 29, no. 1 (2016): 20, https://doi.org/10.1111/johs.12124.

27 Kyle Whyte, "Too Late for Indigenous Climate Justice: Ecological and Relational Tipping Points," *WIREs Climate Change* 11, no. 1 (2020), https://doi.org/10.1002/wcc.603.

28 Ashlee Cunsolo Willox, "Climate Change as the Work of Mourning," *Ethics & the Environment* 17, no. 2 (2012), https://doi.org/10.2979/ethicsenviro.17.2.137.

29 Hannah Knox, "Thinking Like a Climate," *Distinktion: Journal of Social Theory* 16, no. 1 (2015): 103, https://doi.org/10.1080/1600910X.2015.1022565.

30 Bawaka Country et al., "Gathering of the Clouds: Attending to Indigenous Understandings of Time and Climate through Songspirals."

31 Jade Sasser, "Population, Climate Change, and the Embodiment of Environmental Crisis," in *Systemic Crises of Global Climate Change: Intersections of Race, Class and Gender*, eds. Phoebe Godfrey and Denise Torres (Milton Park and New York: Taylor & Francis, 2016), 58.

32 Neimanis and Walker, "Weathering: Climate Change and the "Thick Time" of Transcorporeality," 558.

33 Haraway, *When Species Meet*.

34 Danielle Celermajer, *Summertime: Reflections on a Vanishing Future* (Penguin Random House Australia, 2021).

35 This does run the risk of conflating distinct variations of colonialism and would not be appropriate in all contexts.

36 This does not mean that the book has no relevance for people who are more vulnerable, and less responsible, for changing the climate. It may, but in many cases, such people already have relationally attuned cultures that live more or less sustainably with the world and it is unlikely I will have much insight regarding their lives, needs and experiences. It is also worth noting here that not everyone has the opportunity to live with climate

change, because climate change kills people, typically those marginalised and oppressed by the very systems that warm the planet. Therefore, this book focuses on those of us who are embedded, and have some power, within climate-changing systems because we have the responsibility and capacity to change these systems; it is us who need to transform ourselves.

37 Kari Marie Norgaard, *Living in Denial: Climate Change, Emotions, and Everyday Life* (Cambridge and London: MIT Press, 2011), 61. (Original emphasis)

38 Susan Clayton and Christie Manning, *Psychology and Climate Change: Human Perceptions, Impacts, and Responses* (London, San Diego, Cambridge and Oxford: Academic Press, 2018); Leslie Davenport, *Emotional Resiliency in the Era of Climate Change: A Clinician's Guide* (London and Philidelphia: Jessica Kingsley Publishers, 2017).

39 Geert van Oldenborgh et al., "Attribution of the Australian Bushfire Risk to Anthropogenic Climate Change," *Natural Hazards Earth System Science*, 21 (2021), https://doi.org/10.5194/nhess-2020-69.

40 Blanche Verlie, "'Climatic-Affective Atmospheres': A Conceptual Tool for Affective Scholarship in a Changing Climate," *Emotion, Space and Society* 33 (2019), https://doi.org/10.1016/j.emospa.2019.100623; Anderson, "Affective Atmospheres."

41 Rosemary Randall, "Loss and Climate Change: The Cost of Parallel Narratives," *Ecopsychology* 1, no. 3 (2009), https://doi.org/10.1089/eco.2 009.0034; Cunsolo Willox, "Climate Change as the Work of Mourning."

42 Thom van Dooren, Eben Kirksey, and Ursula Münster, "Multispecies Studies: Cultivating Arts of Attentiveness," *Environmental Humanities* 8, no. 1 (2016): 12, https://doi.org/10.1215/22011919-3527695.

43 For a lengthier exploration of the methodology underpinning this book, see Blanche Verlie, "Encountering, Witnessing and Storying Climate Change's Affective Atmospheres," in *Affective Entanglements: Learning to Live-with Climate Change* (Monash University: Doctor of Philosophy, 2019).

44 David Rousell, Amy Cutter-Mackenzie, and Jasmyne Foster, "Children of an Earth to Come: Speculative Fiction, Geophilosophy and Climate Change Education Research," *Educational Studies* 53, no. 6 (2017): 661, https://doi.org/10.1080/00131946.2017.1369086.

45 For useful personal guidance to help navigate climate anxiety see Sarah Jaquette Ray, *A Field Guide to Climate Anxiety: How to Keep Your Cool on a Warming Planet* (Oakland: University of California Press, 2020); Anouchka Grose, *A Guide to Eco-Anxiety: How to Protect the Planet and Your Mental Health* (London: Watkins Media, 2020).

46 Neimanis and Walker, "Weathering: Climate Change and the "Thick Time" of Transcorporeality."

47 Sefanit Habtom and Megan Scribe, "To Breathe Together: Co-Conspirators for Decolonial Futures," Yellowhead Institute (June 2 2020). https://yellowheadinstitute.org/2020/06/02/to-breathe-together/; Kyle Whyte, "Indigenous Science (Fiction) for the Anthropocene: Ancestral Dystopias and Fantasies of Climate Change Crises," *Environment and Planning E: Nature and Space* 1, no. 1–2 (2018), https://doi.org/10.1177/2514848618777621.

48 For more discussion on such strategies, see Blanche Verlie et al., "Educators' Experiences and Strategies for Responding to Ecological Distress," *Australian Journal of Environmental Education*, Online First (2020), https://doi.org/10.1017/aee.2020.34.

References

Adey, Peter. "Air/Atmospheres of the Megacity." *Theory, Culture & Society* 30, no. 7–8 (2013): 291–308. https://doi.org/10.1177/0263276413501541.

Alaimo, Stacy. "Trans-Corporeal Feminisms and the Ethical Space of Nature." In Material Feminisms, edited by Stacy Alaimo and Susan Hekman, 237–64. Bloomington: Indiana University Press, 2008.

Allam, Lorena, and Carly Earl. "For Centuries the Rivers Sustained Aboriginal Culture. Now They Are Dry, Elders Despair." *The Guardian*, January 22, 2019. https://www.theguardian.com/australia-news/2019/jan/22/murray-darling-river-aboriginal-culture-dry-elders-despair-walgett.

Anderson, Ben. "Affective Atmospheres." *Emotion, Space and Society* 2, no. 2 (2009): 77–81. https://doi.org/10.1016/j.emospa.2009.08.005.

Barad, Karen. *Meeting the Universe Halfway: Quantum Physics and the Entanglement of Matter and Meaning.* Durham and London: Duke University Press, 2007.

Bawaka Country, S. Wright, S. Suchet-Pearson, K. Lloyd, L. Burarrwanga, R. Ganambarr, M. Ganambarr-Stubbs, B. Ganambarr, and D. Maymuru. "Gathering of the Clouds: Attending to Indigenous Understandings of Time and Climate through Songspirals." *Geoforum* 108 (2020): 295–304. https://doi.org/10.1016/j.geoforum.2019.05.017.

Braidotti, Rosi. *The Posthuman.* Cambridge and Malden: Polity Press, 2013.

Celermajer, Danielle. *Summertime: Reflections on a Vanishing Future.* Penguin Random House Australia, 2021.

Celermajer, Danielle, David Schlosberg, Lauren Rickards, Makere Stewart-Harawira, Mathias Thaler, Petra Tschakert, Blanche Verlie, and Christine Winter. "Multispecies Justice: Theories, Challenges, and a Research Agenda for Environmental Politics." *Environmental Politics* 30, no. 1–2 (2020): 119–40. https://doi.org/10.1080/09644016.2020.1827608.

Clayton, Susan, and Christie Manning. *Psychology and Climate Change: Human Perceptions, Impacts, and Responses.* London, San Diego, Cambridge and Oxford: Academic Press, 2018.

Cole, Peter. "Education in an Era of Climate Change: Conversing with Ten Thousand Voices." *TCI (Transnational Curriculum Inquiry)* 13, no. 1 (2016): 3–13.

Cunsolo Willox, Ashlee. "Climate Change as the Work of Mourning." *Ethics & the Environment* 17, no. 2 (2012): 137–64. https://doi.org/10.2979/ethicsenviro.17.2.137.

Davenport, Leslie. *Emotional Resiliency in the Era of Climate Change: A Clinician's Guide.* London and Philidelphia: Jessica Kingsley Publishers, 2017.

Grose, Anouchka. *A Guide to Eco-Anxiety: How to Protect the Planet and Your Mental Health.* London: Watkins Media, 2020.

Habtom, Sefanit, and Megan Scribe. "To Breathe Together: Co-Conspirators for Decolonial Futures." Yellowhead Institute June 2, 2020. https://yellowheadinstitute.org/2020/06/02/to-breathe-together/.

Haltinner, Kristin, and Sarathchandra Dilshani. "Climate Change Skepticism as a Psychological Coping Strategy." *Sociology Compass* 12, no. 6 (2018): 1–10. https://doi.org/10.1111/soc4.12586.

Haraway, Donna. "Situated Knowledges: The Science Question in Feminism and the Privilege of Partial Perspective." *Feminist Studies* 14, no. 3 (1988): 575–99. https://doi.org/10.2307/3178066.

Haraway, Donna. *Staying with the Trouble: Making Kin in the Chthulucene.* Durham and London: Duke University Press, 2016.

Haraway, Donna. *When Species Meet.* Minneapolis: University of Minnesota Press, 2008.

Hulme, Mike. "Climate." *Environmental Humanities* 6 (2015): 175–78.

Hulme, Mike. *Weathered: Cultures of Climate.* Los Angeles, London, New Delhi, Singapore, Washington, Melbourne: SAGE Publications, 2017.

Huntley, Rebecca. *How to Talk About Climate Change in a Way That Makes a Difference.* Crows Nest and London: Murdoch Books, 2020.

Ingold, Tim. *The Life of Lines.* Milton Park and New York: Routledge, 2015.

Knox, Hannah. "Thinking Like a Climate." *Distinktion: Journal of Social Theory* 16, no. 1 (2015): 91–109. https://doi.org/10.1080/1600910X.2015.1022565.

McCormack, Derek "Engineering Affective Atmospheres on the Moving Geographies of the 1897 Andrée Expedition." *Cultural Geographies* 15, no. 4 (2008): 413–30. https://doi.org/10.1177/1474474008094314.

Morton, Timothy. "The Hurricane in My Backyard." *The Atlantic*, 2018. https://www.theatlantic.com/technology/archive/2018/07/the-hurricane-in-my-backyard/564554/.

Moser, Susanne. "Getting Real About It: Meeting the Psychological and Social Demands of a World in Distress." In *Environmental Leadership*, edited by Deborah Gallagher, 900–8. Los Angeles, London, New Delhi, Singapore, Washington DC: SAGE Publications, 2012.

Neimanis, Astrida, and Jennifer Mae Hamilton. "Weathering." *Feminist Review* 118, no. 1 (2018): 80–4. https://doi.org/10.1057/s41305-018-0097-8.

Neimanis, Astrida, and Rachel Loewen Walker. "Weathering: Climate Change and the "Thick Time" of Transcorporeality." *Hypatia* 29, no. 3 (2014): 558–75. https://doi.org/10.1111/hypa.12064.

Norgaard, Kari Marie. *Living in Denial: Climate Change, Emotions, and Everyday Life.* Cambridge and London: MIT Press, 2011.

Plumwood, Val. *Feminism and the Mastery of Nature.* London: Routledge, 1993.

Randall, Rosemary. "Loss and Climate Change: The Cost of Parallel Narratives." *Ecopsychology* 1, no. 3 (2009): 118–29. https://doi.org/10.1089/eco.2009.0034.

Ray, Sarah Jaquette. *A Field Guide to Climate Anxiety: How to Keep Your Cool on a Warming Planet.* Oakland: University of California Press, 2020.

Rickards, Lauren. "Critiquing, Mining and Engaging Anthropocene Science." *Dialogues in Human Geography* 5, no. 3 (2015): 337–42. https://doi.org/10.11 77/2043820615613263.

Rousell, David, Amy Cutter-Mackenzie, and Jasmyne Foster. "Children of an Earth to Come: Speculative Fiction, Geophilosophy and Climate Change Education Research." *Educational Studies* 53, no. 6 (2017): 654–69. https://doi.org/10.1080/00131946.2017.1369086.

Sasser, Jade. "Population, Climate Change, and the Embodiment of Environmental Crisis." In *Systemic Crises of Global Climate Change: Intersections of Race, Class and Gender*, edited by Phoebe Godfrey and Denise Torres. Milton Park and New York: Taylor & Francis, 2016.

Stewart, Kathleen. "Atmospheric Attunements." *Environment and Planning D: Society and Space* 29, no. 3 (2011): 445–53. https://doi.org/10.1068/d9109.

Taylor, Affrica. "Beyond Stewardship: Common World Pedagogies for the Anthropocene." *Environmental Education Research* 23, no. 10 (2017): 1448–61. https://doi.org/10.1080/13504622.2017.1325452.

Todd, Zoe. "An Indigenous Feminist's Take on the Ontological Turn: 'Ontology' Is Just Another Word for Colonialism." *Journal of Historical Sociology* 29, no. 1 (2016): 4–22. https://doi.org/10.1111/johs.12124.

Tuana, Nancy. "Viscous Porosity: Witnessing Katrina." In *Material Feminisms*, edited by Stacy Alaimo and Susan Hekman, 188–213. Bloomington: Indiana University Press, 2008.

van Dooren, Thom, Eben Kirksey, and Ursula Münster. "Multispecies Studies: Cultivating Arts of Attentiveness." *Environmental Humanities* 8, no. 1 (2016): 1–23. https://doi.org/10.1215/22011919-3527695.

van Oldenborgh, Geert, Folmer Krikken, Sophie Lewis, Nicholas Leach, Flavio Lehner, Kate Saunders, Michiel van Weele, et al. "Attribution of the Australian Bushfire Risk to Anthropogenic Climate Change." *Natural Hazards Earth System Science* 21 (2021): 941–60. https://doi.org/10.5194/nhess-2020-69.

Verlie, Blanche. "'Climatic-Affective Atmospheres': A Conceptual Tool for Affective Scholarship in a Changing Climate." *Emotion, Space and Society* 33 (2019): 100623. https://doi.org/10.1016/j.emospa.2019.100623.

Verlie, Blanche. "Encountering, Witnessing and Storying Climate Change's Affective Atmospheres." In *Affective Entanglements: Learning to Live-with Climate Change*. Monash University, Doctor of Philosophy, 2019, 94–125.

Verlie, Blanche. "Rethinking Climate Education: Climate as Entanglement." *Educational Studies* 53, no. 6 (2017): 560–72. https://doi.org/10.1080/00131 946.2017.1357555.

Verlie, Blanche, Emily Clark, Tamara Jarrett, and Emma Supriyono. "Educators' Experiences and Strategies for Responding to Ecological Distress." *Australian Journal of Environmental Education*, Online First (2020): 1–15. https://doi.org/10.1017/aee.2020.34.

Whyte, Kyle. "Indigenous Science (Fiction) for the Anthropocene: Ancestral Dystopias and Fantasies of Climate Change Crises." *Environment and Planning E: Nature and Space* 1, no. 1–2 (2018): 224–42. https://doi.org/1 0.1177/2514848618777621.

Whyte, Kyle. "Too Late for Indigenous Climate Justice: Ecological and Relational Tipping Points." *WIREs Climate Change* 11, no. 1 (2020): e603. https://doi.org/10.1002/wcc.603.

Wright, Alexis. "Deep Weather." *Meanjin* 2 (2011): 70–82. https:// meanjin.com.au/essays/deep-weather/.

2 Feeling the climate crisis

During 2019 and 2020 Australia burned and suffocated for months on end. This was our worst fire season ever, in many regards. From September 2019 to early March 2020 fires burned out of control. During December, more than 420,000 fires were detected across the country.[1] Fires merged with each other, forming 'megafires' of hundreds of thousands of hectares which left firefighters facing fronts of up to 6,000 kilometres. Across the nation, an estimated 24–40 million hectares burned, an area of land equal to or greater than the entire United Kingdom.[2] Over the New Year's holiday period an area half the size of Belgium was evacuated in Gippsland, east Victoria. Thirty-three human people died in the fires, and an estimated three billion vertebrates were killed or displaced.[3]

The fires brought terror, panic and grief, and our political leaders' indifference and ineptitude regarding the fires sparked outrage throughout the nation. Whether on social media or watching the news, footage of firestorms, burnt koalas, fleeing kangaroos and distraught people filled our screens. In addition, the smoke would not let up. Throughout the summer, all of Australia's eastern capital cities, from Brisbane to Hobart, and their regional counterparts, experienced significant smoke pollution that affected over 80% of Australians (see e.g. Figure 2.1). Studies found that the poor air quality led to nearly 5,000 hospital admissions and over 400 human deaths.[4] Depression and anxiety were widely attributed to bushfires and the smoke,[5] with even the *Australian Financial Review* quoting Dante's *Inferno* to describe the atmosphere.[6] Incredibly jarring and disorienting experiences characterised the period, too. I found the footage of a magpie mimicking fire truck sirens deeply haunting, and others described it as 'practically dystopian'[7] and 'the bleakest thing you'll see today.'[8]

Australia's previously unprecedented summers of extreme heat, crippling drought and mass animal deaths had been dubbed 'angry

DOI: 10.4324/9780367441265-2

Figure 2.1 Thick bushfire smoke along the Hume Highway, January 5, 2020. The smoke was this heavy for almost the entirety of my 850-kilometre trip from Bendigo (Central Victoria) to Sydney (New South Wales), as it had been when I made the original trip south on December 22, 2019.

summers' and then in 2018/2019 the 'angriest summer.'[9] This meant that there was nowhere left to go with the 'angry' terminology. This period became the Black Summer, referencing the burnt forests, the carpet of ash that rained across the nation, the midday skies, and the mood. These were dark days, figuratively and literally. Historically, our dark days have been *a* day, or close to. The worst in living memory have been Ash Wednesday (1983) and Black Saturday (2009). The shift to a season as the temporal reference for bushfires speaks to the escalation in scale, distribution, intensity and duration of these fires. But even this term fails to accurately represent this: if we are to use the European-imposed calendar, fires burned from winter all the way to the following autumn.

The 2019/2020 fires were a pertinent, if gut-wrenching, example of the *affectivity* of atmospheres and climates. There is a lot to gain from such an understanding, and as the naming of the Black Summer demonstrates, an appreciation of climate's affective capacities is latent within even the most ecologically insensitive cultures. For example, in English it is common to use the terms 'atmosphere' and 'climate' as metaphors for the emotional, political and/or interpersonal conditions, relations and patterns of a space, group, time or event. With a bit of careful articulation we can amplify this and unearth otherwise suppressed relations, knowledges, motivations and capacities.

In academic literature, the concept of 'affective atmospheres' offers significant steps in this regard. This scholarship theorises affects as embodied, visceral, sensual and somatic energies or forces. Affects can consolidate into emotions and can emerge from them, but they encompass far more than emotions. Emotions are familiar, identifiable and internalised feelings which are experienced by individuated people. By contrast, affects are distributed enigmatic intensities that might be perplexing, fleeting or disorienting, and they often exceed our capacity to make sense of them.[10] Despite their significant influence on us, sometimes we might not even notice them, perhaps because they are so mundane and everyday that we take them for granted. Understanding affects to be atmospheric emphasises that affects form and move in ways similar to meteorological phenomena: they emerge, condense, drift and dissipate according to their relations with other parts of the world.[11] Just like air, affects swirl, waft, interpermeate and exude from human and non-human bodies and practices, and in so doing, they effect change in the world. In this sense, affective atmospheres help us attune to the embodied relationships between beings, and how and why they change and morph.[12]

Australia's 2019/2020 fires demonstrate that not only are affects atmospheric, but that climatic phenomena are affective.[13] Climate affects us, but equally, as the Australian Government's adoration of fossil fuels indicates, our desires and affective norms propel the ways we participate in climate.[14] *Climate is feeling.* Or to be a little more precise, climate is distributed and dynamic patterns of affect; the climate crisis is a radical disruption in those patterns; and climate action requires affective transformation. The embodied, more-than-human relations that compose climate, and which we live-with as part of climate, all have affective – energetic, emotional – dimensions.

As I explore throughout this book, attuning to humans' affective entanglement with the climatic has significant implications for how we approach human-climate relations, and more specifically, climate change engagement. This chapter demonstrates that human affective experiences of climatic phenomena emerge from, and contribute to, our embodied relations with the more-than-human world. In order to illustrate this, I explore the political, cultural, scientific, emotional and embodied entanglements that coalesced to create the 'end of days'[15] feeling of the fires, and how this atmosphere kindled a nascent collective climate consciousness. I am not exploring, at least not as the central focus, people's lived experiences of being directly in the fire zones, or of losing loved ones or homes. These experiences deserve far more space than I can give here.[16] Rather, I am focusing on the larger population's experiences of and responses to the bushfire smoke and mediated representations of the fires which unfolded in a context where the majority of Australians knew people or places that could have been or were directly affected, and where the smoke and media brought and maintained their attention to that. This is in order to focus on the multiple ways through which atmospheres – as climatic and affective – are composed: through immediate gaseous envelopment, cultural representations, and multi-temporal and geographically dispersed inter-personal and more-than-human relations. Such an approach foregrounds the ways that climates, feelings, and people re-compose each other through complex and ongoing entanglements. This theorisation of the co-creation of gaseous conditions and human subjectivities (our sense of self) informs the remainder of the book which discusses the practices through which we can cultivate careful climate collaboration in the midst of planetary catastrophe.

During mid-2020 as I wrote this chapter, the Arctic reached temperatures of 100 °F (38 °C) and experienced significant wildfires. Shortly after, California recorded the world's hottest ever temperature (130°F/ 54.4°C) and experienced off-the-charts 'gigafires' that produced apocalyptic skies all too reminiscent of Australia's.[17] Such increasing fire conditions around the globe have led some to suggest we are entering the age of the Pyrocene.[18] Articulated with devastating accuracy by Greta Thunberg's refrain 'our house is on fire,' the combustion of fossil fuels is returning to us as the incineration of our homes. While the discussion in this chapter is specifically about Australia's most recent fires (at the time of writing), unfortunately, similar experiences are likely to be ignited around the world in coming years.

***

The smoky skies of the 2019/2020 fires are evidence of our corporeal entanglement with/in our more-than-human relations. The ashes floating through the skies were the 'muted embodiments' of dead, and in some cases now possibly extinct, beings who were 'no longer individuated or recognisable, and yet all there...in a dispersed, dusty mass.'[19] In breathing the smoke, we inhaled incinerated forests and wildlife, and the tiny particles of their charred bodies made their way into our lungs, our blood, our organs and our brains. This 'macabre atmospheric communion'[20] produced a 'thick fear' that 'settled in so many of our hearts.'[21] The smoke's capacity to diffuse and permeate all kinds of bodies epitomised Stacy Alaimo's concept of transcorporeality which refers to how 'all creatures, as embodied beings, are intermeshed with the dynamic, material world, which crosses through them, transforms them, and is transformed by them.'[22]

Efforts at constructing borders to counteract this transcorporeal pollution proliferated as people tried to reduce the human health risks and avoid the general disgust of the smoke. Hardware stores sold out of P2 face masks, although they turned out to be basically ineffective anyway. Health agencies advised us to stay at home and close the doors and windows. Interior windowsills dusted with ash led to the rolling up of wet towels to fashion makeshift draught stoppers. The smoke was as uncontrollable as the fires though, and it wafted through crevices with ease, disabling human agencies. One sufferer of rheumatoid arthritis described the 'huge loss of bodily autonomy' that the heavy smoke induced, noting it 'massively' increased her levels of inflammation and joint pain and that this got worse every time she took a breath.[23] Urban fire brigades were overwhelmed by smoke alarms triggered by the high levels of indoor pollution, and evacuation protocols that directed people outside into the even worse air highlighted how unanticipated and unmanageable this situation was. Other technologies were also overwhelmed or disrupted by the smoke: MRIs stopped working, and in a cruel twist, the smoke appeared as rain on the Bureau of Meteorology's radars.[24]

Climate change provides important lessons here about the incapacity of humans to manage atmospheric phenomena. The fires were found to be 30% more likely because of fossil-fuelled climate change,[25] affirming the ability of humans to affect the planet's atmospheres. But the bushfires doubled Australia's annual greenhouse gas emissions.[26] This is what climate scientists call a positive feedback loop, when fossil-fuelled climate change catalyses ecological processes that further warm the globe. In such circumstances, whatever ability humans had to limit global heating become overwhelmed by the planet's own potential to change its climate.[27] These additional greenhouse gas emissions were produced because fires burned 'out of control' for months, despite a highly developed country's best efforts to 'contain' them. Rather than controlling the climate, the fires demonstrate that we are instead interfering with and unleashing climate's own agency.[28]

The smoke's unavoidable intrusions into our bodies also elicited broader emotional distress. For those of us in the inner cities, the visceral ways we encountered the smoke ensured that even if we wanted to, we could not escape the knowledge of the harrowing realities faced by rural and regional communities, including our non-human companions. Throughout the fire season, empathic relations were cultivated as we were 'feeling other people's feelings in the air,'[29] or to put it more bluntly, we were immersed in their extermination. Wildlife carers spoke of their grief while caring for injured animals, made all the more overwhelming as there was 'the scent of death in every breath' in the smoky skies enveloping them.[30] Over time I noticed that this odour of obliteration changed depending on the proximity of the fires and wind direction: volatile eucalyptus like a campfire, which was disconcertingly nostalgic; dirty like pollution, which made me feel deeply resigned; stale like ashtrays, which was the most physically vile. The skies also changed colour: white, grey, dusty yellow and a Mad Max–like orange within Sydney where I was living; and scarlet red and midnight black in areas most immediately under threat of the out-of-control fires. As the skies reddened, so did our eyes and the secretions from our noses. The inflammation in our throats was a niggling reminder of the fate of our forests.

The distress this embodied testimony generated was amplified by mediated representations of the smoke and fires. Satellite images documented the smoke traversing the globe and scientists informed us that toxicity reached more than twenty times the 'hazardous' level in some places. My stomach dropped watching *The Guardian's* animated graph which had to reorient the scale to accommodate these extremes.[31] The unending news feeds of apocalyptic imagery meant that the bushfires were distressing for many, many Australians, not just those who were directly affected. These anxieties were further inflamed by scientific knowledge of climate change. As David Wallace-Wells suggests, in our changing climate, bushfires are 'terrifying and immediate, no matter how far from a fire zone you live. They offer up vivid, scarring images it can be impossible not to read as portents of future nightmares even as they document present tragedies and horrors.'[32] The affective intensification that these multiple forms of knowledge generated is apparent in Charlotte Wood's discussion of her experience in inner-Sydney:

> In news updates about the fires, it's now commonplace to hear two horrific phrases: 'seek shelter' and 'too late to leave'…We watch the footage – those walls of orange flame storeys high – with our hands over our mouths…This is what the scientists have warned us about, *begged* us to think of, all these years. It's here. And it's going to get worse…We have the 'Fires Near Me' app on our

phones now, but I'm careful not to zoom out too far from our immediate 50km zone. If you do, it's easy to panic. There are so many little fire symbols they overlap...And zooming out brings the existential horror of what all this really means...There's nothing like going to sleep with the taste of ash in your throat to give you an actual, physiological understanding of real fear.[33]

The scientific, cultural and embodied knowledges of the fires and smoke all functioned as affective prostheses, enabling us to vicariously experience some of the terror others were suffering, to some extent. But it was the smoke, more than anything else, that rendered apparent how our practices of knowing climate change are enacted through being enmeshed within it. As Dayna Scott suggests, 'there is a boundary transgression inherent in the act of smelling: to become aware of a scent is to have already inhaled it.'[34] This bodily knowledge was augmented with air quality indexes, maps of ember attack zones, photographs of deceased firefighters, videos of blackened koalas and underlying awareness of our collapsing climate. This meant that inhaling our cremated kin was not just a respiratory irritation but a visceral existential crisis. As one writer put it, 'the dread is palpable... we're breathing it in.'[35] The horrifying atmospheres demonstrate that our affective experiences of climate change are not internally generated in our psyches, but emerge from our emplaced and embodied relations with/in planetary processes. Bushfires are an iconic example of this affective entanglement, as they can 'mobilize and transmit latent ecological energies, innervating the connective tissues that exist between interdependent webs of human and other-than-human life.'[36]

Of course, not everyone experienced the 2019/2020 fires in the same way. Being affectively entangled with others does not mean being or feeling the same as them. The affective forces of climate change condense around different bodies in different ways, meaning that not everyone has to weather the same affective conditions.[37] For city dwellers like me, our embodied experiences of the smoke were only a taste – literally – of the terror and suffering experienced by those actually living amongst, and dying because of, the fires. We know that climate change's causes and impacts are distributed in starkly unequal ways, and this is true of the affectivity of climate change.

For example, in countries like Australia, settler aspirations for unfettered economic growth generate deep intimacies between our politicians and fossil fuels,[38] energy sources that are only extractable through the long-running and ongoing exertion of power over First Nations' people.[39] These same racial hierarchies constitute some of the social inequalities that render marginalised peoples more vulnerable to climate change.[40] Building on Sara Ahmed's assertion that racism can feel *like* a storm,[41] the 2019/2020 fires demonstrate that atmospheric phenomena *are* affectively racialised.

Such 'settler atmospherics'[42] were brutally apparent when fires closed in on Lithgow in New South Wales and authorities evacuated everyone except the inmates at the Correctional Centre, almost a quarter of whom are Indigenous (in comparison to the national population of 3%).[43] This fits all too neatly with the systemic 'weaponising' of bad weather in the racist Australian carceral system, practices like maintaining poor ventilation and removing fans that create 'cumulative and deeply embodied' affective violences.[44] More broadly, for many Indigenous Australians the 2019/2020 bushfires led to 'deep psychological wounds' that ruptured their identities because the Country that burned is understood to be 'an extension of self.'[45] This is a distinct kind of ecological grief, and it is a violence that is further amplified by the intergenerational and ongoing 'trauma of dispossession and neglect' that Indigenous Australians experience every day.[46]

Meanwhile, our political leaders appeared to remain largely unaffected by the atmosphere, with Prime Minister Scott Morrison spouting a litany of platitudes throughout the summer. For example, on New Year's Eve an image of 11-year-old Finn Burns, clad in gas mask and enveloped in blood red skies, piloting his family's boat away from the ensuing inferno was beamed around the world. On New Year's Day, Morrison proclaimed that there was 'no better place to raise kids' than Australia.[47] The contrast between the complacent indifference of such privileged climate deniers and the world-shattering terror and grief of others exemplifies the affective climate injustice perpetuated when insulated people turn away from the climate crisis. Those who actively strive to remain unaffected outsource the emotional labour and affective pain of climate change onto others more marginalised, and less complicit, than themselves.

Yet the ways that political leaders gaslit the public meant that the fires did not only incinerate the forest, they inflamed human spirits and fired up political tensions. During the fires, the emotional climate denial of leading politicians[48] catalysed an emergent collective climate consciousness in the Australian public. The smoke also played a significant role in sparking these affective affinities. Climate protests are always atmospheric, in the sense of being affectively influenced by the weather: chanting 'climate action now' when corralled under a baking sun leaves you feeling both exhausted and justified. But the stark novelty of the ashen skies made this entanglement of the meteorological and the affective all the more apparent. Across the country, rallies throughout December and January protested not just *about* climate change and the fires, but *in* the smoke. News media interviewed people who said that they had never protested before but that they were absolutely seething about the smoke and thus decided to attend. One commented that he never thought he would 'have to rally for my right to breathe.'[49]

Protest signs referencing the suffocating conditions proliferated at rallies. Prime Minister Scott Morrison's nickname ScoMo was appropriated into SmoKo. Having celebrated his recent electoral victory with the phrase 'how good is Australia!?,' placards at protests challenged: 'how good is breathing?' Another in central Sydney proclaimed:

They BURN

We CHOKE

You DO NOTHING

One sign had images of Sydney which they compared to Mordor, and another with a picture of the planet proclaimed 'her lungs hurt too.' The protests were full of P2 facemasks adorned with 'climate action now' or 'climate emergency.'[50] Perhaps the most on-point placard I saw consisted of just one small remaining corner of charred cardboard. On December 19, 2019, Australia's second hottest day ever (beaten only by the day before),[51] school strikers led a protest outside the Prime Minister's official residence to campaign for him to come home from holiday in Hawaii and *do something* about the fires and climate change. At this protest, 13-year-old Izzy Raj-Seppings was moved on by the police under threat of force. As Izzy recounted, the context and stake of the protests, the solidarity of her fellow protestors, the 'blazing sun,' swirling smoke and riot squad combined to create a 'whirlwind of emotions.'[52]

The atmospheres of the 2019/2020 fires were so affective that they traversed the globe, just as the smoke circumnavigated it. Some ex-pats felt 'on edge – all the time' because of the constant stream of media updates about the fires; others drew on this distress to 'get political,' get together, and to write to politicians.[53] One report from a protest in London demonstrates that the fires elicited care, action and visceral affective responses, far from the site of the local crisis: 'the seemingly endless footage of Australia's bushfire crisis…had motivated plenty of newcomers' and at this particular protest, 'a funereal tone radiated' amongst the protestors.[54]

As pyrocumulonimbus clouds filled skies across south-eastern Australia, the public mood darkened, and yet from these apocalyptic atmospheres a climate concerned collectivity emerged, even if only temporarily. The fires' atmospheres swirled around and through us, and were viscerally felt, connecting us in shared-but-differentiated experiences. Gamilaraay and Yawalaraay reporter Lorena Allam reflected on the summer, and hoped: 'maybe this summer is the turning point, where our collective grief turns to action and we recognise the knowledge that First Nations people want to share, to make sure these horrors are never repeated.'[55] As if responding to Allam's call, settler Australian Alicia Flynn reflected:

> This planetary moment of crisis, despair, sorrow and fear, also feels like the closest we have come to a sense of collective entanglement on such a vast scale – emotionally, materially and existentially...Aboriginal people, like many different First Nations people around the world, have known the end of times, the end of worlds and yet their knowledge persists...This crisis might just help us finally surrender our savage settler ego and, along with the sorrow and deep ongoing sorry work that needs to happen, support First Peoples to fully recuperate the knowledge of caring for this place in the ways that they have cultivated for millennia.[56]

Discussions of Indigenous Australians' practice of cultural burning also received a surge of media attention during and after the fires,[57] described by some as a 'quantum shift in public awareness of Aboriginal fire management.'[58] Some of these discussions perpetuated the inaccurate understanding of cultural burning as a traditional form of hazard reduction which seeks to protect human life and property. However, conversations also stretched beyond this to discuss the complexities and nuances of cultural burning which is a relational kinship-based practice that cares for Country. Through cool burns tailored to each specific place, cultural burning allows for 'the involvement of other than human beings such as bettongs, bandicoots, lyrebirds, wombats and brush turkeys who all assist with cultural burning by turning over and reducing the leaf litter.'[59] In addition to numerous interviews with cultural burning experts Oliver Costello and Victor Steffensen,[60] discussions also acknowledged the risks of cultural appropriation and the need to return sovereignty to Indigenous Australians – not only to prevent escalating bushfire threats but to heal

Country and decolonise more broadly.[61] This suggests that perhaps, as Allam hoped, climatic traumas such as Australia's 2019/2020 fires can spark cultural transformation away from colonial extractivism and towards more ecologically and socially just relations.

On the other hand, climatic tragedies can also compel a resurgence of the very ideologies that set them in motion. Under the guise of urgency, quick fixes that perpetuate hierarchical domination can proliferate. Throughout and following the fire season, approaches calling for a 'war-like' response to the climate crisis, including the suppression of democracy, increased in volume and frequency.[62] Personally, I was immensely relieved when HMAS Choules was deployed to evacuate some of the 4,000 people who were stranded on the beach in remote Mallacoota over New Year's Eve, yet a militarised response to climate change more broadly will not contribute to climate justice.[63] As Tony Birch argues, while we need to speed up some elements of our response, we need to slow down in other ways.[64] The 2019/2020 fires were an opportunity to rethink extractive modes of being and glimmers of this occurred. However, it would be naive to assume that this happened in any systematic, significant or enduring way.

***

Australia's 2019/2020 bushfires demonstrate the intense affectivity of climatic phenomena and provide important insight about how this unfolds and what it can do. The emotional and embodied intensities of climate change can emerge from both immediate unfolding disasters and awareness of slower global trends. The ferocious fires and the chronic smoke pollution of 2019/2020 intersected with underlying eco-anxiety and long-running systemic climate denial to generate widespread panic, grief and anger. These atmospheres were composed by political, cultural, geographical, scientific, interpersonal and more-than-human forces and relations. The multiscalar energetic dispersals of the 2019/2020 fires affected people's physiological wellbeing, experiences of place, beliefs about the future, interpersonal relations, political alignments and position in power hierarchies. These relational, embodied, more-than-human affective forces contributed to people's emerging sense of self, perception of reality and understanding of how the world works.

For the climate conscious, the fires were experienced as 'the death of the planet come to visit you in your home.'[65] It brought 'a fear that it won't be over soon, or ever.'[66] The reality that periods like this are becoming a 'new normal' which will in turn soon be superseded is almost unthinkable. This chapter has not touched on the world-shattering experiences of those in the fire zones, either during or following the fires. Yet even focusing on the broader public's experiences of the smoke and mediated representations of the fires demonstrates the overwhelming distress that climate change can elicit, especially when local disasters are situated within awareness of the projected global possibilities. As Fiona Wright articulates, the 2019/2020 fires were 'the aftermath of a catastrophe' and 'the shape of things to come, and it's hard to hold both of these things together.'[67]

But as climate science tells us, we are going to be living with the climate crisis, in one way or another, from here on. Because of past emissions and the climate's complexity, the heat latent in the climate system will continue warming the planet even if we magically halted all emissions now.[68] The 2019/2020 fires' loss of three billion vertebrates[69] occurred at just one degree of global heating. It is going to take 'rapid and far-reaching transitions' in energy, land, infrastructure and industrial systems to prevent exceeding 1.5 degrees.[70] Few nations are making progress anywhere in line with this.

Given how unfathomable this is, we have a society-wide educational challenge on our hands: we need to *learn* to live with climate change. The following chapters discuss three affective practices that will be key in this journey: encountering, witnessing, and storying climate change. These practices are enacted *with* climate change, and were unfolding during the 2019/2020 fires, albeit in largely unorganised and unconscious ways: we inhaled choking smoke, we beheld previously unimaginable fires, and we produced shocked narratives of this. With more explicit attention to how these practices work and what they can do, we will be better placed to support people to engage in them in promising ways. The remainder of this book seeks to provide such guidance.

Notes

1 Niko Kommenda and Josh Holder, "Visual Guide: See How Australia's Bushfires Are Raging across the Country," *The Guardian*, January 7, 2020. https://www.theguardian.com/australia-news/2020/jan/07/visual-guide-see-how-australias-bushfires-are-raging-across-the-country.

2 Royal Commission into National Natural Disaster Arrangements, *Interim Observations 31 August 2020*, Commonwealth of Australia (2020), https://naturaldisaster.royalcommission.gov.au/publications/interim-observations-31-august-2020.

3 WWF, *Australia's 2019–2020 Bushfires: The Wildlife Toll*, World Wildlife Fund (2020), https://www.wwf.org.au/news/news/2020/3-billion-animals-impacted-by-australia-bushfire-crisis#gs.p9wksk.
4 Fay Johnston et al., "Unprecedented Health Costs of Smoke-Related PM2.5 from the 2019–20 Australian Megafires," *Nature Sustainability* 4 (2020), https://doi.org/10.1038/s41893-020-00610-5; Georgia Hitch, "Bushfire Royal Commission Hears That Black Summer Smoke Killed Nearly 450 People," *ABC News*, May 26, 2020, https://www.abc.net.au/news/2020-05-26/bushfire-royal-commission-hearings-smoke-killed-445-people/12286094.
5 Ella Oar, Mel Taylor, and Fran Molloy, "Eco-Anxiety Climbs as Fires, Smoke and Animal Deaths Trigger Fear and Trauma," *The Lighthouse*, January 9, 2020, https://lighthouse.mq.edu.au/article/january-2020/eco-anxiety-climbs-as-fires,-smoke-and-animal-deaths-trigger-fear-and-trauma
6 Andrew Clark, "Fires Spread a Pall of Depression," *Australian Financial Review*, December 13, 2019, https://www.afr.com/policy/energy-and-climate/fires-spread-a-pall-of-depression-20191212-p53jg7.
7 Andy Moser, "Watch This Australian Magpie Perfectly Mimic the Sound of Emergency Sirens," *Mashable*, January 3, 2020, https://mashable.com/article/australian-magpie-mimic-siren-sound-wildfires/.
8 James Felton, "This Australian Magpie Mimicking Emergency Sirens Is the Bleakest Thing You'll See Today," *IFL Science*, January 3, 2020, https://www.iflscience.com/plants-and-animals/this-australian-magpie-mimicking-emergency-sirens-is-the-bleakest-thing-youll-see-today/.
9 Will Steffen et al., *Angry Summer 2016/17: Climate Change Super-Charging Extreme Weather*, Climate Council (2017), https://www.climatecouncil.org.au/resources/angry-summer-report/; Will Steffen et al., *The Angriest Summer*, The Climate Council, 2019, https://www.climatecouncil.org.au/resources/angriest-summer/.
10 Gerda Roelvink and Magdalena Zolkos, "Affective Ontologies: Post-Humanist Perspectives on the Self, Feeling and Intersubjectivity," *Emotion, Space and Society* 14 (2015), https://doi.org/10.1016/j.emospa.2014.07.003; Pieter Vermeulen, "Posthuman Affect," *European Journal of English Studies* 18, no. 2 (2014), https://doi.org/10.1080/13825577.2014.917001.
11 Ben Anderson, "Affective Atmospheres," *Emotion, Space and Society* 2, no. 2 (2009), https://doi.org/10.1016/j.emospa.2009.08.005.
12 Peter Adey, "Air/Atmospheres of the Megacity," *Theory, Culture & Society* 30, no. 7–8 (2013), https://doi.org/10.1177/0263276413501541; Michael Buser, "Thinking through Nonrepresentational and Affective Atmospheres in Planning Theory and Practice," *Planning Theory* 13, no. 3 (2013), https://doi.org/10.1177/1473095213491744.
13 Blanche Verlie, "'Climatic-Affective Atmospheres': A Conceptual Tool for Affective Scholarship in a Changing Climate," *Emotion, Space and Society* 33 (2019), https://doi.org/10.1016/j.emospa.2019.100623. Susanne Gannon, "Ordinary Atmospheres and Minor Weather Events," *Departures in Critical Qualitative Research* 5, no. 4 (2016), https://doi.org/10.1525/dcqr.2016.5.4.79, http://dcqr.ucpress.edu/content/5/4/79.abstract; Gail Adams-Hutcheson, "Farming in the Troposphere: Drawing Together Affective Atmospheres and Elemental Geographies," *Social & Cultural Geography* 20, no. 7 (2019), https://doi.org/10.1080/14649365.2017.1406982.

14 Blanche Verlie, "Greenhouse Gaslighting: Scott Morrison's Emotional Manipulation from Climate Apathy to Fake Empathy," *Sydney Environment Institute*, January 14, 2020, https://www.sydney.edu.au/environment-institute/blog/greenhouse-gaslighting-scott-morrisons-emotional-manipulation-from-climate-apathy-to-fake-empathy/. Kari Marie Norgaard, *Living in Denial: Climate Change, Emotions, and Everyday Life* (Cambridge and London: MIT Press, 2011).

15 Naaman Zhou and Josh Taylor, "The Big Smoke: How Bushfires Cast a Pall over the Australian Summer," *The Guardian*, December 5, 2019, https://www.theguardian.com/australia-news/2019/dec/05/the-big-smoke-how-bushfires-cast-a-pall-over-the-australian-summer.

16 For detailed first hand accounts of the fires, see Michael Rowland, *Black Summer: Stories of Loss, Courage and Community from the 2019–2020 Bushfires* (ABC Books, 2021). Danielle Celermajer, *Summertime: Reflections on a Vanishing Future* (Penguin Random House Australia, 2021).

17 David Wallace-Wells, "California Has Australian Problems Now," *Intelligencer*, August 21, 2020, https://nymag.com/intelligencer/2020/08/climate-change-has-led-to-extreme-wildfires-in-california.html.

18 Stephen Pyne, "Big Fire; or Introducing the Pyrocene," *Fire* 1, no. 1 (2018). https://doi.org/10.3390/fire1010001.

19 Sophie Chao, "A World of Ashes," *Sydney Environment Institute*, January 8, 2020, http://sydney.edu.au/environment-institute/opinion/a-world-of-ashes/.

20 Chao, "A World of Ashes."

21 Fiona Wright, "An Air of Dread," *Kill Your Darlings*, December 12, 2019, https://www.killyourdarlings.com.au/article/an-air-of-dread-the-mental-toll-of-sydneys-bushfire-smoke/.

22 Stacy Alaimo, "Trans-Corporeality," in *Posthuman Glossary*, eds. Rosi Braidotti and Maria Hlavajova (Bloomsbury Publishing, 2018), 435.

23 The Guardian, "The Frontline. Inside Australia's Climate Emergency: The Air We Breathe," *The Guardian*, February 20, 2020, https://www.theguardian.com/environment/ng-interactive/2020/feb/20/the-toxic-air-we-breathe-the-health-crisis-from-australias-bushfires.

24 Ben Graham, "'It's Not Rain': Eerie Bom Map Explained," *Daily Mercury*, December 6, 2019, https://m.dailymercury.com.au/news/its-not-rain-eerie-bom-map-explained/3896100/.

25 Geert van Oldenborgh et al., "Attribution of the Australian Bushfire Risk to Anthropogenic Climate Change," *Natural Hazards Earth System Science* 21 (2020), https://doi.org/10.5194/nhess-2020-69.

26 Andrew Freedman, "Australia's Greenhouse Gas Emissions Effectively Double as a Result of Unprecedented Bush Fires," *The Washington Post*, January 25, 2020, https://www.washingtonpost.com/weather/2020/01/24/australia-bush-fires-have-nearly-doubled-countrys-annual-greenhouse-gas-emissions/.

27 Will Steffen et al., "Trajectories of the Earth System in the Anthropocene," *Proceedings of the National Academy of Sciences* 115, no. 33 (2018), https://doi.org/10.1073/pnas.1810141115.

28 Nigel Clark, "Volatile Worlds, Vulnerable Bodies," *Theory, Culture & Society* 27, no. 2–3 (2010), https://doi.org/10.1177/0263276409356000.

29 Dora Zhang, "Notes on Atmosphere," *Qui Parle* 27, no. 1 (2018): 126, https://doi.org/10.1215/10418385-4383010.
30 Calla Wahlquist, "'Brutal Business': Bushfire Devastation Causes 'Collective Grief' among Wildlife Carers," *The Guardian*, January 17, 2020, https://www.theguardian.com/australia-news/2020/jan/17/brutal-business-bushfire-devastation-causes-collective-grief-among-wildlife-carers.
31 Kommenda and Holder, "Visual Guide: See How Australia's Bushfires Are Raging across the Country."
32 Wallace-Wells, "California Has Australian Problems Now."
33 Charlotte Wood, "From Disbelief to Dread: The Dismal New Routine of Life in Sydney's Smoke Haze," *The Guardian*, December 7, 2019, https://www.theguardian.com/australia-news/2019/dec/07/from-disbelief-to-dread-the-dismal-new-routine-of-life-in-sydneys-smoke-haze. (Original emphasis).
34 Dayna Nadine Scott, "'We Are the Monitors Now': Experiential Knowledge, Transcorporeality and Environmental Justice," *Social & Legal Studies* 25, no. 3 (2016): 27, https://doi.org/10.1177/0964663915601166.
35 Wright, "An Air of Dread."
36 Jobb Arnold, "Feeling the Fires of Climate Change: Land Affect in Canada's Tar Sands," in *Affective Ecocriticism: Emotion, Embodiment, Environment*, eds. Kyle Bladow and Jennifer Ladino, (University of Nebraska Press, 2018), 197.
37 Astrida Neimanis and Jennifer Mae Hamilton, "Weathering," *Feminist Review* 118, no. 1 (2018), https://doi.org/10.1057/s41305-018-0097-8.
38 Cara Daggett, "Petro-Masculinity: Fossil Fuels and Authoritarian Desire," *Millennium* 47, no. 1 (2018), https://doi.org/10.1177/0305829818775817.
39 Tony Birch, "Climate Change, Mining and Traditional Indigenous Knowledge in Australia," *Social Inclusion* 4, no. 1 (2016), https://doi.org/http://dx.doi.org/10.17645/si.v4i1.442; Kyle Whyte, "Indigenous Climate Change Studies: Indigenizing Futures, Decolonizing the Anthropocene," *English Language Notes* 55, no. 1 (2017), https://www.muse.jhu.edu/article/711473.
40 Nancy Tuana, "Climate Apartheid: The Forgetting of Race in the Anthropocene," *Critical Philosophy of Race* 7, no. 1 (2019), https://doi.org/10.5325/critphilrace.7.1.0001; Anna Kaijser and Annica Kronsell, "Climate Change through the Lens of Intersectionality," *Environmental Politics* 23, no. 3 (2014), https://doi.org/10.1080/09644016.2013.835203.
41 Sara Ahmed, "Atmospheric Walls," *Feminist Killjoys*, September 15, 2014, https://feministkilljoys.com/2014/09/15/atmospheric-walls/.
42 Kristen Simmons, "Settler Atmospherics," *Member Voices, Fieldsights*, November 20, 2017, https://culanth.org/fieldsights/settler-atmospherics.
43 Madeline Hayman-Reber, "Government Defends Decision to Keep Lithgow Prisoners inside as Fire Rages," *NITV News*, December 23, 2019, https://www.sbs.com.au/nitv/article/2019/12/23/government-defends-decision-keep-lithgow-prisoners-inside-fire-rages1.
44 Stella Maynard, "Weaponised Weathers: Heat, Don Dale and 'Everything-Ist' Prison Abolition," *Sydney Environment Institute*, August 12, 2019, https://sei.sydney.edu.au/opinion/weaponised-weathers-heat-don-dale-everything-ist-prison-abolition/.
45 Liz Cameron cited in Miya Yamanouchi, "Aboriginal People in Mourning over Bushfires: 'We Grieve Because Country Is Wounded and Bleeding,'"

Sarajevo Times, January 29, 2020, https://www.sarajevotimes.com/aboriginal-people-in-mourning-over-bushfires-we-grieve-because-country-is-wounded-and-bleeding/.

46 Bhiamie Williamson, Jessica Weir, and Vanessa Cavanagh, "Strength from Perpetual Grief: How Aboriginal People Experience the Bushfire Crisis," *The Conversation*, January 10, 2020, https://theconversation.com/strength-from-perpetual-grief-how-aboriginal-people-experience-the-bushfire-crisis-129448.

47 Amy Remeikis, "'No Better Place to Raise Kids': Scott Morrison's New Year Message to a Burning Australia," *The Guardian*, January 1, 2020, https://www.theguardian.com/australia-news/2020/jan/01/no-better-place-to-raise-kids-scott-morrison-new-year-message-burning-australia.

48 Verlie, "Greenhouse Gaslighting: Scott Morrison's Emotional Manipulation from Climate Apathy to Fake Empathy."

49 SBS News, "As Bushfire Smoke Choked NSW, Sydneysiders Rallied to Demand Climate Action," *SBS News*, December 11, 2019, https://www.sbs.com.au/news/as-bushfire-smoke-choked-nsw-sydneysiders-rallied-to-demand-climate-action. A comment the protestor may hopefully have reflected on following the resurgence of Black Lives Matter protests that erupted in response to the murder by suffocation of George Floyd by Minneapolis police in May 2020.

50 Following the onset of the COVID-19 pandemic, seeing Australians wearing face masks on mass no longer seems strange, but at the time, this was both rare and eerie.

51 On Thursday the 19th of December 2019, Australia had an average maximum temperature of 41.0°C (105.8°F), beaten only by the day before which clocked in at 41.9°C (107.2°F)

52 Isolde (Izzy) Raj-Seppings, "I'm the 13-Year-Old Police Threatened to Arrest at the Kirribilli House Protest. This Is Why I Did It," *The Guardian*, December 21, 2019, https://www.theguardian.com/australia-news/commentisfree/2019/dec/21/im-the-13-year-old-police-threatened-to-arrest-at-the-kirribilli-house-protest-this-is-why-i-did-it

53 Stephanie Jane Capper, "As the Bushfire Crisis Unfolded, Australian Expats Ran the Gamut from Guilt to Grief," *ABC News*, February 4, 2020, https://www.abc.net.au/news/2020-02-04/australia-bushfire-crisis-expats-felt-guilty-grief-helplessness/11917052.

54 Bevan Shields, "'Wake up and Smell the Smoke': Bushfire Protests Spread to London," *The Sydney Morning Herald*, January 11, 2020, https://www.smh.com.au/world/europe/wake-up-and-smell-the-smoke-bushfire-protests-spread-to-london-20200111-p53qkg.html.

55 Lorena Allam, "For First Nations People the Bushfires Bring a Particular Grief, Burning What Makes Us Who We Are," *The Guardian*, January 6, 2020, https://www.theguardian.com/commentisfree/2020/jan/06/for-first-nations-people-the-bushfires-bring-a-particular-grief-burning-what-makes-us-who-we-are.

56 Alicia Flynn, "To Ged Kearney: These Fires Feel Like the End," *FireFeels*, January 14, 2020, https://firefeels.org/2020/01/14/to-ged-kearney-these-fires-feel-like-the-end/.

57 For example, see the following international media pieces: Gary Nunn, "Australia Fires: Aboriginal Planners Say the Bush 'Needs to Burn',"

BBC, January 12, 2020, https://www.bbc.com/news/world-australia-51043 828. Aarti Betigeri, "How Australia's Indigenous Experts Could Help Deal with Devastating Wildfires," *Time*, January 14, 2020, https://time.com/5 764521/australia-bushfires-indigenous-fire-practices/.

58 David Bowman, Greg Lehman, and Andry Sculthorpe, "Australia, You Have Unfinished Business. It's Time to Let Our 'Fire People' Care for This Land," *The Conversation*, May 28, 2020, https://theconversation.com/ australia-you-have-unfinished-business-its-time-to-let-our-fire-people-care-for-this-land-135196.

59 Shaun Hooper, "Cultural Burning Is About More Than Just Hazard Reduction," *IndigenousX*, January 7, 2020, https://indigenousx.com.au/ cultural-burning-is-about-more-than-just-hazard-reduction/. See also Victor Steffensen, *Fire Country: How Indigenous Fire Management Could Help Save Australia* (Melbourne and Sydney: Hardie Grant Travel, 2020).

60 For example: Susan Chenery and Ben Cheshire, "Fighting Fire with Fire: Passed on through the Generations, Could Indigenous Cultural Burning Save Australia's Landscape from Another Catastrophic Bushfire Season?," *ABC News*, April 13, 2020, https://www.abc.net.au/news/2020-04-13/how-victor-steffensen-is-fighting-fire-with-fire/11866478?nw=0; Isabella Higgins, "Indigenous Fire Practices Have Been Used to Quell Bushfires for Thousands of Years, Experts Say," *ABC News*, January 9, 2020, https:// www.abc.net.au/news/2020-01-09/indigenous-cultural-fire-burning-method-has-benefits-experts-say/11853096; Oliver Costello, "Our Ancestors Managed Fire Country for Millennia. We Yearn to Burn Once More," *The Guardian*, January 28, 2020, https://www.theguardian.com/commentisfree/2 020/jan/28/our-ancestors-managed-fire-country-for-millennia-we-yearn-to-burn-once-more.

61 Timothy Neale, "What Are Whitefellas Talking About When We Talk About "Cultural Burning"?," *Inside Story*, April 17, 2020, https://insidestory.org.au/ what-are-whitefellas-talking-about-when-we-talk-about-cultural-burning/.

62 Sarah Martin, "Labor MP Urges War-Like National Mobilisation to Tackle Australia's Existential Threat of Climate Crisis," *The Guardian*, January 6, 2020, https://www.theguardian.com/australia-news/2020/jan/06/labor-mp-urges-war-like-national-mobilisation-to-tackle-australias-existential-threat-of-climate-crisis; Mike Foley, "'Climate Change War' Demands National, Military-Style Response: Ex-Fire Chiefs," *The Sydney Morning Herald*, July 6, 2020, https://www.smh.com.au/politics/federal/climate-change-war-demands-national-military-style-response-ex-fire-chiefs-20200706-p559h7.html; Marie McInerney, "Seven Snapshots from the National Climate Emergency Summit," *Croakey*, February 19, 2020, https://croakey.org/seven-snapshots-from-the-national-climate-emergency-summit/; Marie McInerney and Peter Garrett, "Peter Garrett Calls for Australia to Go onto a 'War' Footing on Climate Change," *Croakey*, February 18, 2020, https://www.croakey.org/ peter-garrett-calls-for-australia-to-go-onto-a-war-footing-on-climate-change/.

63 Kelly Albion, "Why Calling for a 'Climate Emergency' Is Not Climate Justice," Australian Youth Climate Coalition (2019). https:// www.aycc.org.au/climatejustice_not_climateemergency.

64 Tony Birch, "Climate Change, Recognition and Social Place-Making," in *Unstable Relations: Indigenous People and Environmentalism in Contemporary*

Australia, eds. Eve Vincent and Timothy Neale (University of Western Australia Publishing, 2016).

65 First Dog on the Moon, "Living through Endless Weeks of Dirty Air – It Does Your Head in and Your Lungs," *The Guardian*, December 5, 2019, https://www.theguardian.com/commentisfree/2019/dec/04/living-through-endless-weeks-of-dirty-air-it-does-your-head-in-and-your-lungs.

66 Wood, "From Disbelief to Dread: The Dismal New Routine of Life in Sydney's Smoke Haze."

67 Wright, "An Air of Dread."

68 IPCC, *Climate Change 2001: Synthesis Report of the Third Assessment Report*, Intergovernmental Panel on Climate Change (2001), 88–90. https://www.ipcc.ch/site/assets/uploads/2018/03/q1to9-1.pdf; Richard Rood, "If We Stopped Emitting Greenhouse Gases Right Now, Would We Stop Climate Change?," *The Conversation*, July 5, 2017, https://theconversation.com/if-we-stopped-emitting-greenhouse-gases-right-now-would-we-stop-climate-change-78882.

69 WWF, *Australia's 2019–2020 Bushfires: The Wildlife Toll*.

70 IPCC, *Global Warming of 1.5°C: Summary for Policy Makers*, Intergovernmental Panel on Climate Change (2018), 15, http://www.ipcc.ch/report/sr15/.

References

Adams-Hutcheson, Gail. "Farming in the Troposphere: Drawing Together Affective Atmospheres and Elemental Geographies." *Social & Cultural Geography* 20, no. 7 (2019): 1004–23. https://doi.org/10.1080/14649365.2017.1406982.

Adey, Peter. "Air/Atmospheres of the Megacity." *Theory, Culture & Society* 30, no. 7–8 (2013): 291–308. https://doi.org/10.1177/0263276413501541.

Ahmed, Sara. "Atmospheric Walls." *Feminist Killjoys*, September 15, 2014. https://feministkilljoys.com/2014/09/15/atmospheric-walls/.

Alaimo, Stacy. "Trans-Corporeality." In *Posthuman Glossary*, edited by Rosi Braidotti and Maria Hlavajova, 435–38. Bloomsbury Publishing, 2018.

Albion, Kelly. "Why Calling for a 'Climate Emergency' Is Not Climate Justice." (2019). Australian Youth Climate Coalition, https://www.aycc.org.au/climatejustice_not_climateemergency.

Allam, Lorena. "For First Nations People the Bushfires Bring a Particular Grief, Burning What Makes Us Who We Are." *The Guardian*, January 6, 2020. https://www.theguardian.com/commentisfree/2020/jan/06/for-first-nations-people-the-bushfires-bring-a-particular-grief-burning-what-makes-us-who-we-are.

Anderson, Ben. "Affective Atmospheres." *Emotion, Space and Society* 2, no. 2 (2009): 77–81. https://doi.org/10.1016/j.emospa.2009.08.005.

Arnold, Jobb. "Feeling the Fires of Climate Change: Land Affect in Canada's Tar Sands." In *Affective Ecocriticism: Emotion, Embodiment, Environment*, edited by Kyle Bladow and Jennifer Ladino, 95–116. University of Nebraska Press, 2018.

Betigeri, Aarti. "How Australia's Indigenous Experts Could Help Deal with Devastating Wildfires." *Time*, January 14, 2020. https://time.com/5764521/australia-bushfires-indigenous-fire-practices/.

Birch, Tony. "Climate Change, Mining and Traditional Indigenous Knowledge in Australia." *Social Inclusion* 4, no. 1 (2016): 92–101. https://doi.org/http://dx.doi.org/10.17645/si.v4i1.442.

Birch, Tony. "Climate Change, Recognition and Social Place-Making." In *Unstable Relations: Indigenous People and Environmentalism in Contemporary Australia*, edited by Eve Vincent and Timothy Neale, 356–83: University of Western Australia Publishing, 2016.

Bowman, David, Greg Lehman, and Andry Sculthorpe. "Australia, You Have Unfinished Business. It's Time to Let Our 'Fire People' Care for This Land." *The Conversation*, May 28, 2020. https://theconversation.com/australia-you-have-unfinished-business-its-time-to-let-our-fire-people-care-for-this-land-135196.

Buser, Michael. "Thinking through Nonrepresentational and Affective Atmospheres in Planning Theory and Practice." *Planning Theory* 13, no. 3 (2013): 227–43. https://doi.org/10.1177/1473095213491744.

Capper, Stephanie Jane. "As the Bushfire Crisis Unfolded, Australian Expats Ran the Gamut from Guilt to Grief." *ABC News*, February 4, 2020. https://www.abc.net.au/news/2020-02-04/australia-bushfire-crisis-expats-felt-guilty-grief-helplessness/11917052.

Celermajer, Danielle. *Summertime: Reflections on a Vanishing Future*. Penguin Random House Australia, 2021.

Chao, Sophie. "A World of Ashes." *Sydney Environment Institute*, January 8, 2020. http://sydney.edu.au/environment-institute/opinion/a-world-of-ashes/.

Chenery, Susan, and Ben Cheshire. "Fighting Fire with Fire: Passed on through the Generations, Could Indigenous Cultural Burning Save Australia's Landscape from Another Catastrophic Bushfire Season?" *ABC News*, April 13, 2020. https://www.abc.net.au/news/2020-04-13/how-victor-steffensen-is-fighting-fire-with-fire/11866478?nw=0.

Clark, Andrew. "Fires Spread a Pall of Depression." *Australian Financial Review*, December 13, 2019. https://www.afr.com/policy/energy-and-climate/fires-spread-a-pall-of-depression-20191212-p53jg7.

Clark, Nigel. "Volatile Worlds, Vulnerable Bodies." *Theory, Culture & Society* 27, no. 2–3 (2010): 31–53. https://doi.org/10.1177/0263276409356000.

Costello, Oliver. "Our Ancestors Managed Fire Country for Millennia. We Yearn to Burn Once More." *The Guardian*, January 28, 2020. https://www.theguardian.com/commentisfree/2020/jan/28/our-ancestors-managed-fire-country-for-millennia-we-yearn-to-burn-once-more.

Daggett, Cara. "Petro-Masculinity: Fossil Fuels and Authoritarian Desire." *Millennium* 47, no. 1 (2018): 25–44. https://doi.org/10.1177/0305829818775817.

Felton, James. "This Australian Magpie Mimicking Emergency Sirens Is the Bleakest Thing You'll See Today." *IFL Science*, January 3, 2020. https://www.iflscience.com/plants-and-animals/this-australian-magpie-mimicking-emergency-sirens-is-the-bleakest-thing-youll-see-today/.

First Dog on the Moon. "Living through Endless Weeks of Dirty Air – It Does Your Head in and Your Lungs." *The Guardian*, December 5, 2019. https://www.theguardian.com/commentisfree/2019/dec/04/living-through-endless-weeks-of-dirty-air-it-does-your-head-in-and-your-lungs.

Flynn, Alicia. "To Ged Kearney: These Fires Feel Like the End." *FireFeels*, January 14, 2020. https://firefeels.org/2020/01/14/to-ged-kearney-these-fires-feel-like-the-end/.

Foley, Mike. "'Climate Change War' Demands National, Military-Style Response: Ex-Fire Chiefs." *The Sydney Morning Herald*, July 6, 2020. https://www.smh.com.au/politics/federal/climate-change-war-demands-national-military-style-response-ex-fire-chiefs-20200706-p559h7.html.

Freedman, Andrew. "Australia's Greenhouse Gas Emissions Effectively Double as a Result of Unprecedented Bush Fires." *The Washington Post*, January 25, 2020. https://www.washingtonpost.com/weather/2020/01/24/australia-bush-fires-have-nearly-doubled-countrys-annual-greenhouse-gas-emissions/.

Gannon, Susanne. "Ordinary Atmospheres and Minor Weather Events." *Departures in Critical Qualitative Research* 5, no. 4 (2016): 79–90. https://doi.org/10.1525/dcqr.2016.5.4.79. http://dcqr.ucpress.edu/content/5/4/79.abstract.

Graham, Ben. "'It's Not Rain': Eerie Bom Map Explained." *Daily Mercury*, December 6, 2019. https://m.dailymercury.com.au/news/its-not-rain-eerie-bom-map-explained/3896100/.

Hayman-Reber, Madeline, "Government Defends Decision to Keep Lithgow Prisoners inside as Fire Rages," *NITV News*, December 23, 2019. https://www.sbs.com.au/nitv/article/2019/12/23/government-defends-decision-keep-lithgow-prisoners-inside-fire-rages1.

Higgins, Isabella. "Indigenous Fire Practices Have Been Used to Quell Bushfires for Thousands of Years, Experts Say." *ABC News*, January 9, 2020. https://www.abc.net.au/news/2020-01-09/indigenous-cultural-fire-burning-method-has-benefits-experts-say/11853096.

Hitch, Georgia. "Bushfire Royal Commission Hears That Black Summer Smoke Killed Nearly 450 People." *ABC News*, May 26, 2020. https://www.abc.net.au/news/2020-05-26/bushfire-royal-commission-hearings-smoke-killed-445-people/12286094.

Hooper, Shaun. "Cultural Burning Is About More Than Just Hazard Reduction." *IndigenousX*, January 7, 2020. https://indigenousx.com.au/cultural-burning-is-about-more-than-just-hazard-reduction/.

IPCC. *Climate Change 2001: Synthesis Report of the Third Assessment Report*. Intergovernmental Panel on Climate Change (2001). https://www.ipcc.ch/site/assets/uploads/2018/03/q1to9-1.pdf.

IPCC. *Global Warming of 1.5°C: Summary for Policy Makers.* Intergovernmental Panel on Climate Change (2018). http://www.ipcc.ch/report/sr15/.

Johnston, Fay, Nicolas Borchers-Arriagada, Geoffrey Morgan, Bin Jalaludin, Andrew Palmer, Grant Williamson, and David Bowman. "Unprecedented Health Costs of Smoke-Related PM2.5 from the 2019–20 Australian Megafires." *Nature Sustainability* 4 (2020): 42–7. https://doi.org/10.1038/s41893-020-00610-5.

Kaijser, Anna, and Annica Kronsell. "Climate Change through the Lens of Intersectionality." *Environmental Politics* 23, no. 3 (2014): 417–33. https://doi.org/10.1080/09644016.2013.835203.

Kommenda, Niko, and Josh Holder. "Visual Guide: See How Australia's Bushfires Are Raging across the Country." *The Guardian*, January 7, 2020. https://www.theguardian.com/australia-news/2020/jan/07/visual-guide-see-how-australias-bushfires-are-raging-across-the-country.

Martin, Sarah. "Labor MP Urges War-Like National Mobilisation to Tackle Australia's Existential Threat of Climate Crisis." *The Guardian*, January 6, 2020. https://www.theguardian.com/australia-news/2020/jan/06/labor-mp-urges-war-like-national-mobilisation-to-tackle-australias-existential-threat-of-climate-crisis.

Maynard, Stella, "Weaponised Weathers: Heat, Don Dale and 'Everything-ist' Prison Abolition," *Sydney Environment Institute*, August 12, 2019. https://sei.sydney.edu.au/opinion/weaponised-weathers-heat-don-dale-everything-ist-prison-abolition/.

McInerney, Marie. "Seven Snapshots from the National Climate Emergency Summit." *Croakey*, February 19, 2020. https://croakey.org/seven-snapshots-from-the-national-climate-emergency-summit/.

McInerney, Marie, and Peter Garrett. "Peter Garrett Calls for Australia to Go onto a 'War' Footing on Climate Change." *Croakey*, February 18, 2020. https://www.croakey.org/peter-garrett-calls-for-australia-to-go-onto-a-war-footing-on-climate-change/.

Moser, Andy. "Watch This Australian Magpie Perfectly Mimic the Sound of Emergency Sirens." *Mashable*, January 3, 2020. https://mashable.com/article/australian-magpie-mimic-siren-sound-wildfires/.

Neale, Timothy. "What Are Whitefellas Talking About When We Talk About 'Cultural Burning'?" *Inside Story*, April 17, 2020. https://insidestory.org.au/what-are-whitefellas-talking-about-when-we-talk-about-cultural-burning/.

Neimanis, Astrida, and Jennifer Mae Hamilton. "Weathering." *Feminist Review* 118, no. 1 (2018): 80–4. https://doi.org/10.1057/s41305-018-0097-8.

Norgaard, Kari Marie. *Living in Denial: Climate Change, Emotions, and Everyday Life.* Cambridge and London: MIT Press, 2011.

Nunn, Gary. "Australia Fires: Aboriginal Planners Say the Bush 'Needs to Burn'." *BBC*, January 12, 2020. https://www.bbc.com/news/world-australia-51043828.

Oar, Ella, Mel Taylor, and Fran Molloy. "Eco-Anxiety Climbs as Fires, Smoke and Animal Deaths Trigger Fear and Trauma." *The Lighthouse*, January 9, 2020. https://lighthouse.mq.edu.au/article/january-2020/eco-anxiety-climbs-as-fires,-smoke-and-animal-deaths-trigger-fear-and-trauma.

Pyne, Stephen. "Big Fire; or Introducing the Pyrocene." *Fire* 1, no. 1 (2018): 1–3. https://doi.org/10.3390/fire1010001.

Raj-Seppings, Isolde (Izzy). "I'm the 13-Year-Old Police Threatened to Arrest at the Kirribilli House Protest. This Is Why I Did It." *The Guardian*, December 21, 2019. https://www.theguardian.com/australia-news/commentisfree/2019/dec/21/im-the-13-year-old-police-threatened-to-arrest-at-the-kirribilli-house-protest-this-is-why-i-did-it.

Remeikis, Amy. "'No Better Place to Raise Kids': Scott Morrison's New Year Message to a Burning Australia." *The Guardian*, January 1, 2020. https://www.theguardian.com/australia-news/2020/jan/01/no-better-place-to-raise-kids-scott-morrison-new-year-message-burning-australia.

Roelvink, Gerda, and Magdalena Zolkos. "Affective Ontologies: Post-Humanist Perspectives on the Self, Feeling and Intersubjectivity." *Emotion, Space and Society* 14 (2015): 47–9. https://doi.org/10.1016/j.emospa.2014.07.003.

Rood, Richard. "If We Stopped Emitting Greenhouse Gases Right Now, Would We Stop Climate Change?" *The Conversation*, July 5, 2017. https://theconversation.com/if-we-stopped-emitting-greenhouse-gases-right-now-would-we-stop-climate-change-78882.

Rowland, Michael. *Black Summer: Stories of Loss, Courage and Community from the 2019–2020 Bushfires*. ABC Books, 2021.

Royal Commission into National Natural Disaster Arrangements. *Interim Observations 31 August 2020*. Commonwealth of Australia (2020). https://naturaldisaster.royalcommission.gov.au/publications/interim-observations-31-august-2020.

SBS News. "As Bushfire Smoke Choked NSW, Sydneysiders Rallied to Demand Climate Action." *SBS News*, December 11, 2019. https://www.sbs.com.au/news/as-bushfire-smoke-choked-nsw-sydneysiders-rallied-to-demand-climate-action.

Scott, Dayna Nadine. "'We Are the Monitors Now' Experiential Knowledge, Transcorporeality and Environmental Justice." *Social & Legal Studies* 25, no. 3 (2016): 261–87. https://doi.org/10.1177/0964663915601166.

Shields, Bevan. "'Wake up and Smell the Smoke': Bushfire Protests Spread to London." *The Sydney Morning Herald*, January 11, 2020. https://www.smh.com.au/world/europe/wake-up-and-smell-the-smoke-bushfire-protests-spread-to-london-20200111-p53qkg.html.

Simmons, Kristen. "Settler Atmospherics." *Member Voices, Fieldsights*, November 20, 2017. https://culanth.org/fieldsights/settler-atmospherics.

Steffen, Will, Annika Dean, Martin Rice, and Greg Mullins. *The Angriest Summer*. The Climate Council, 2019. https://doi.org/10.3390/fire1010001.

Steffen, Will, Johan Rockström, Katherine Richardson, Timothy M. Lenton, Carl Folke, Diana Liverman, Colin P. Summerhayes, *et al.* "Trajectories of the Earth System in the Anthropocene." *Proceedings of the National Academy of Sciences* 115, no. 33 (2018): 8252–59. https://doi.org/10.1073/pnas.1810141115.

Steffen, Will, Andrew Stock, David Alexander, and Martin Rice. *Angry Summer 2016/17: Climate Change Super-Charging Extreme Weather.* Climate Council, 2017. https://www.climatecouncil.org.au/resources/angry-summer-report/.

Steffensen, Victor. *Fire Country: How Indigenous Fire Management Could Help Save Australia.* Melbourne and Sydney: Hardie Grant Travel, 2020.

The Guardian. "The Frontline. Inside Australia's Climate Emergency: The Air We Breathe." *The Guardian*, February 20, 2020. https://www.theguardian.com/environment/ng-interactive/2020/feb/20/the-toxic-air-we-breathe-the-health-crisis-from-australias-bushfires.

Tuana, Nancy. "Climate Apartheid: The Forgetting of Race in the Anthropocene." *Critical Philosophy of Race* 7, no. 1 (2019): 1–31. https://doi.org/10.5325/critphilrace.7.1.0001

van Oldenborgh, Geert, Folmer Krikken, Sophie Lewis, Nicholas Leach, Flavio Lehner, Kate Saunders, Michiel van Weele, *et al.* "Attribution of the Australian Bushfire Risk to Anthropogenic Climate Change." *Natural Hazards Earth System Science* 21 (2021): 941–60. https://doi.org/10.5194/nhess-2020-69.

Verlie, Blanche. "'Climatic-Affective Atmospheres': A Conceptual Tool for Affective Scholarship in a Changing Climate." *Emotion, Space and Society* 33 (2019): 100623. https://doi.org/10.1016/j.emospa.2019.100623.

Verlie, Blanche. "Greenhouse Gaslighting: Scott Morrison's Emotional Manipulation from Climate Apathy to Fake Empathy." *Sydney Environment Institute*, January 14, 2020. https://www.sydney.edu.au/environment-institute/blog/greenhouse-gaslighting-scott-morrisons-emotional-manipulation-from-climate-apathy-to-fake-empathy/.

Vermeulen, Pieter. "Posthuman Affect." *European Journal of English Studies* 18, no. 2 (2014): 121–34. https://doi.org/10.1080/13825577.2014.917001.

Wahlquist, Calla. "'Brutal Business': Bushfire Devastation Causes 'Collective Grief' among Wildlife Carers." *The Guardian*, January 17, 2020. https://www.theguardian.com/australia-news/2020/jan/17/brutal-business-bushfire-devastation-causes-collective-grief-among-wildlife-carers.

Wallace-Wells, David. "California Has Australian Problems Now." *Intelligencer*, August 21, 2020. https://nymag.com/intelligencer/2020/08/climate-change-has-led-to-extreme-wildfires-in-california.html.

Whyte, Kyle. "Indigenous Climate Change Studies: Indigenizing Futures, Decolonizing the Anthropocene." *English Language Notes* 55, no. 1 (2017): 153–62. https://www.muse.jhu.edu/article/711473.

Williamson, Bhiamie, Jessica Weir, and Vanessa Cavanagh. "Strength from Perpetual Grief: How Aboriginal People Experience the Bushfire Crisis." *The Conversation*, January 10, 2020. https://theconversation.com/strength-from-perpetual-grief-how-aboriginal-people-experience-the-bushfire-crisis-129448.

Wood, Charlotte. "From Disbelief to Dread: The Dismal New Routine of Life in Sydney's Smoke Haze." *The Guardian*, December 7, 2019. https://www.theguardian.com/australia-news/2019/dec/07/from-disbelief-to-dread-the-dismal-new-routine-of-life-in-sydneys-smoke-haze.

Wright, Fiona. "An Air of Dread." *Kill Your Darlings*, December 12, 2019. https://www.killyourdarlings.com.au/article/an-air-of-dread-the-mental-toll-of-sydneys-bushfire-smoke/.

WWF. *Australia's 2019–2020 Bushfires: The Wildlife Toll*. World Wildlife Fund, 2020. https://www.wwf.org.au/news/news/2020/3-billion-animals-impacted-by-australia-bushfire-crisis#gs.p9wksk.

Yamanouchi, Miya. "Aboriginal People in Mourning over Bushfires: 'We Grieve Because Country Is Wounded and Bleeding'." *Sarajevo Times*, January 29, 2020. https://www.sarajevotimes.com/aboriginal-people-in-mourning-over-bushfires-we-grieve-because-country-is-wounded-and-bleeding/.

Zhang, Dora. "Notes on Atmosphere." *Qui Parle* 27, no. 1 (2018): 121–55. https://doi.org/10.1215/10418385-4383010.

Zhou, Naaman, and Josh Taylor. "The Big Smoke: How Bushfires Cast a Pall over the Australian Summer." *The Guardian*, December 5, 2019. https://www.theguardian.com/australia-news/2019/dec/05/the-big-smoke-how-bushfires-cast-a-pall-over-the-australian-summer.

3 Encountering climate anxiety

The first thing I think of when I think of climate change, is that I don't want to think about it.

Perhaps this is because much of the airplay surrounding climate change follows a worst-case-scenario narrative. It is a constant reminder that the Earth is fucked. Climate change is everything and it is going to affect everything. It is a massive issue globally, it is the biggest issue that we as custodians of this planet will ever have to contend with.

Learning about climate change broadened my mind on the eventualities to come: privileged countries are able to inadvertently ruin the lives of the deprived. The increased floods in Dhaka, the severity of drought arising in Cape Town, the extreme unpredictability of weather occurring in Toronto and the increased likelihood of hurricanes decimating coastal regions represents a future that becomes more real year after year: mass migration, mass extinction, conflict, hunger, desertification, destruction, loss. I don't think we even grasp the complexity and nuance behind such blunt projections. I think that it will be in the details that the real misery is experienced.

As I live in a high-carbon country and benefit from all the luxuries it brings, I too am in some way a denier.

Realising this came at a great emotional cost to myself. I questioned whether my own actions were creating any positive change. Was I simply another facet at the root of this problem? The hypocrisy played on my conscience, and caused me many sleepless nights, as I grappled with the contradictions between what I practice and what I preach.

Am I doing enough?!

Am I actually doing anything useful at all?!

DOI: 10.4324/9780367441265-3

I feel unsettled because I know I have less of a dependence on climate than others purely because I was born in a wealthy country and live in a well-off family.

I feel bitter towards individuals and systems and fail to understand why people are not being charged for climate crimes.

Australians are some of the worst climate sceptics in the world. I feel that people must become numb to hearing about this issue all the time and it just becomes white noise.

That disinterest sums up a lot of the general public's view on climate change. In the real world, who really cares? Who is going to go out of their way to prevent increasing greenhouse gas emissions? I know that climate change is a collective problem, but who is going to actually contribute to a solution?

I thought, if only the media would portray things accurately, because they are to blame. But it is much more complicated than that. Australia's history is built off of coal. In many ways, we still rely on coal. The denial is cultural because our economic prosperity is tangled with it. Many people's identity is intertwined with it. When you think about it like that, like a change in identity, it is easier to see why people are in denial.

A challenge in acquiring weekly knowledge of climate change has been to keep walking the scary line of learning. Week to week there is always a moment where I am filled with sadness at how I, who am so young, can feel so passionate about a situation seemingly moving backwards. What gets me the most is that the societies and the people that contribute most to climate change are the ones that will be least affected. What happened to karma? That just makes me so angry. In fact we are the ones who cause most of these issues and for them to be faced with the consequences?

I've been crying myself to sleep a lot lately. And crying at random times too. It's not as though I watch a video about climate change, and I cry during it. I mean sometimes that happens. It's more like, something little happens, like my toast burns, and I have an existential breakdown because I think it's a metaphor for how the world is burning because we aren't paying attention.

I found myself dry retching in the shower for over an hour one evening. The contractions of my stomach muscles, sense of my throat exploding, and my whole body convulsing, felt like I was trying to spew up some kind of demon, a wretchedness, a loneliness and desperation, a sense of loss for all that could have been but probably won't, for that which is but will no longer be.

It is such an emotional challenge to deal with, especially when you accept the fact that you and the society we live in today are to blame.

***

In contrast to the volatile atmospheres of Australia's 2019/2020 bushfires, my climate change classes, like so many efforts to engage people with climate change, occur in 'climate-controlled' spaces. Nevertheless, as the vignette above explores, climate change infiltrates and indeed is embedded within these 'thermal enclosures'[1] in all manner of ways. Imagery, videos, graphs, statistics and stories presented in course learning resources, the ecologies of the buildings and campus, and students' emplaced climatic relations all contribute to climatic-affective atmospheres[2] infusing these supposedly insulated spaces. Our classrooms are thus inescapably implicated in and imbued with the material and cultural elements of climate change.

Climate change is a trauma of geological proportions that can 'enter into the classroom not only as discursive topics, but as transcorporeal forces that pass through and infect the very structures of bodies.'[3] For example, as students and I discuss the systems that expose society's most marginalised to lethal heat stress, our bodily reactions such as sweaty armpits, flushed cheeks and croaky voices belie the 'thermal monotony'[4] of our air-conditioned comfort. Subsequently, these embodied responses to discussing climate change contribute to the classroom atmospheres, which in turn go on to affect others in these spaces, extending climate change's affective flows. Attuning to the affective experiences of engaging with climate change in such insulated places offers an alternative geography of climate change, where the embodied impacts of the climate crisis can be seemingly divorced from the times, spaces and indeed bodies in which they are normally presumed to be located. The extensile and expansive energetic disruptions of climate change can permeate our bodies. In doing so, they can cultivate 'entangled empathy'[5] through highlighting our somatic connectivity with others, both known, unknown, and not yet in existence.[6]

Engaging with climate change can be inspiring, energising, and invigorating, but very often it is also deeply distressing. Petra

Tschakert and co-authors describe climate change as 'one thousand ways to experience loss,'[7] and of course, these losses are only going to escalate. In this context, Lesley Head writes that grief will be our constant companion, and one we need to learn to live with.[8] For many experiencing ecological distress, it can feel like grief is our only companion, with isolation and loneliness common among those panicking about the state of the planet. Yet, climate anxiety and grief are never a self-contained experience; they are not feelings encased within our psyches or even our skin-bound bodies. Rather, they are transcorporeal experiences which are generated through our affective entanglement with the wider atmospheric world. As I argue in this chapter, they are evidence of climate change's capacity to exceed our control, rupture our worlds and reconfigure our selves.

In many of our efforts to respond to climate change, we seek to increase people's knowledge about climate change. This has obvious value but also comes with risks, depending on the ethos we bring to those pursuits. Too often, we are seeking mastery, domination and control, the very ideals that got us into this mess.[9] We do this when we try to completely comprehend climate change, under the premise that if we could fully and accurately know climate change then we would be (better) able to control the climate.[10] This approach objectifies climate change and encloses it within our minds, leaving little openness for our transcorporeal body-climate relationships to activate alternative ways of knowing and responding to climate change.

Our efforts at climate change engagement must move beyond pedagogies based on humans 'knowing about' climate change. Climate anxiety might provide the opportunity needed to spur such reconsideration, because it undermines the notion that knowing more is a straightforward process of gaining control.[11] Climate anxiety arises when people are aware of and know about climate change – this is the 'scary line of learning' articulated in the vignette above. Knowledge may be power, but when it comes to climate change, it can also be incredibly disabling. The relationship between knowledge of climate change and the unsettling distress this can generate provides an important impetus to rethink the whole endeavour of seeking to know climate change.

Attuning to the processes through which humans encounter climate change can help shake loose of such mentalities of 'climate control.' To encounter is to meet another in an unexpected way, or in a way that is surprising or challenging. Encounters are not preorganised, they do not go to plan, they are not on our terms and we are not in charge of them. Encounters are relational exchanges that reconfigure those who participate in them; they *counter* existing ways of being and relating.[12]

Attuning to the ways that we encounter climate change is to bring an attention to the ways that climate change is disrupting, pushing and reconfiguring us, and how disorienting, bewildering and destabilising that can be. As demonstrated in this chapter's vignette, our practices of producing, interpreting and recalling information about climate change are often processes through which the unruly capacities of climate change exceed our mental faculties and elicit all kinds of embodied, interpersonal and existential unravellings.

We can encounter climate change in virtually unlimited ways. Direct lived experience of extreme weather – such as Australia's 2019/2020 bushfires, discussed in the previous chapter – is the most obvious mode, but we can also encounter climate change through graphs of carbon dioxide, stories of climate injustice, news reports from disaster zones, arguments with sceptical family members, campaigns for carbon pricing policies, changing birdlife in the garden, disrupted public transport, our own nausea when we speak about climate change, school students striking in the street; the list goes on.

All of these diverse modes of encountering climate change can generate climate anxiety, a multifaceted and varied experience that can encompass distressing emotions and a broader existential unmooring.[13] While climate anxiety can intersect with and contribute to clinically diagnosable mental illnesses, it is not itself pathological. Rather, it is a rational, if painful, response to climate change that can motivate action, although of course it can also lead to disengagement and burnout.[14] Climate anxiety, as I refer to it, is more than the short-term concern or worry that can arise when people are confronted with information about climate change. Rather, it is a medium- to long-term sense that their world – their relationships, assumptions, dreams, security and identity – is in the process of ending. This might be a slow dread interspersed with periods of acute panic, but it can manifest in all kinds of ways. While the terminology of eco- and climate 'anxiety' appears to be taking hold, 'ecological distress' more accurately denotes this wide range of emotional and affective experiences and avoids the risk of pathologisation.[15] For example, for people in high-carbon societies, their climate anxiety is often pervaded by guilt (as this chapter's vignette explores), clarifying that climate change becomes an existential threat because it jeopardises people's survival as well as their sense of themselves as morally competent individuals. As the experience of guilt shows, ecological distress is socially differentiated, and our emotional engagements with climate change are a site of climate politics.

As I explore in this chapter (and across the book), when we encounter climate change, we become climate-changed. Focusing specifically on experiences of ecological distress, including anxiety, I attune to what these emotional, affective and existential destabilisations do to (certain) people's sense of self and world. I demonstrate that experiences of ecological distress participate in the articulation, consolidation, and/or decomposition of 'the sense of the "me" or "us" harbouring the emotion.'[16] My main purpose is not to provide an in-depth analysis of any individual person's experiences, as important as that may be, nor to comprehensively document the characteristics of experiences of climate anxiety. Rather, through attuning to the unruly transcorporeality (i.e. embodied, relational, and more-than-humanness) of climate change's affective agency, I seek to show that encountering climate anxiety disrupts the sense of individuality that we so often take for granted, and that it can therefore contribute to the emergence of new ways of being human.

***

Anxiety is one of the most common emotions people feel regarding climate change. Climate change's uncertainty is one of the core drivers of this anxiety, as it undermines Western notions of stability and progress. Scientists confirm that it is 'extremely likely' that industrial practices are heating the planet's atmosphere and that this means it is 'virtually certain' that there will be more frequent hot extremes.[17] However, despite such high-level certainties, the finer details of exactly how climate change will play out are not fully foreseeable. This is largely because we do not know to what extent industrial societies will limit their production of greenhouse gases. But it is also because the climate is composed through all the relations of all the elements of the planet. It is not possible to perfectly predict how these interdependencies will respond to climate change's cascading disruptions and disappearances, even though we have solid knowledge of likely trends. Such 'guaranteed uncertainty' can generate feelings of confusion, worry and anxiety, as this statement from one of my students indicates: 'I feel an uncertain future is inevitable, and it deeply upsets me.'

Such feelings of anxiety are not just psychological but visceral, as another student's reflection indicates:

> [climate change] often develops some sort of anxiety in me, or even just a mild sick feeling, the feeling of not being able to predict the future and predict the best thing to do whilst minimising as much

harm as possible. This feeling often comes by me when I watch how the leaders of Australia approach issues like these, and it makes me feel sick.

Similarly to dry retching (as narrated in this chapter's vignette) and others' raspy panic attacks,[18] such experiences of climate anxiety demonstrate that our human bodies are energetically enmeshed with the planet. Planetary gaseous imbalances can generate metabolic and respiratory agitation, momentary manifestations of epochal disturbances that can be felt in our tummies, throats and chests.[19] Climate anxiety's physiological materialisations demand that we attune to our climatic transcorporeality and resist the tendency to frame climate anxiety as an internal, psychological phenomenon.

In addition, climate anxiety is a deeply political experience that emerges from particular social relations. The previous comment's reference to the loss of intellectual and moral control – generated by government indifference – demonstrates that climate change is not just upsetting, it can be unsettling. Settler-colonial and over-industrialised cultures promise their (privileged) communities certainty, progress and control of the future.[20] We are also told that if we want to be ethical people, we can achieve this through our individual choices. But climate change undermines these narratives of individualised morality and entitlement to a bright future. As youth climate activist Jamie Margolin puts it: 'it doesn't matter how talented we are, how much work we put in, how many dreams we have, the reality is, my generation has been committed to a planet that is collapsing.'[21] This breakdown of the progress narrative constitutes a kind of cultural trauma for industrial societies;[22] as Lesley Head articulates, part of our climate anxiety is an anticipatory grief for the loss of our modern selves and their promising futures.[23] In these ways, climate anxiety destabilises the settler sense of belonging to the future; as school strikers proclaim, their (presumed) future is being stolen. Consider Greta Thunberg's speech from the Houses of Parliament in London, 2019:

> I was fortunate to be born in a time and place where everyone told us to dream big; I could become whatever I wanted to. I could live wherever I wanted to. People like me had everything we needed and more. Things our grandparents could not even dream of. We had everything we could ever wish for and yet now we may have nothing. Now we probably don't even have a future anymore.[24]

All around the world, young people like Greta – privileged *and* vulnerable – are enraged by the anachronistic mythology of eternal economic growth that is decimating the planet they are inheriting, and rightfully so. But Indigenous peoples remind us that colonialism has been ending worlds and stealing futures for centuries.[25] Climate change brings the violences of capitalist ideology home to its heirs, and as the future slips beyond our control, we can become unsettled.[26] These unnerving feelings force us to reckon with the myths of extractive cultures of domination, as Greta continues:

> Because that future was sold so that a small number of people could make unimaginable amounts of money. It was stolen from us every time you said that the sky was the limit, and that you only live once. You lied to us. You gave us false hope. You told us that the future was something to look forward to.[27]

Through its capacity to dislocate privileged people from their presumed bright futures, climate anxiety provides a constructive opportunity to challenge, resist and rework these dominant cultures. If engaged with carefully, these unsettling feelings might open us to decolonial modes of being, although the desire to regain control of the world is a terribly resilient trait of contemporary capitalism. Still, as some of the statements in this chapter's vignette indicate, climatic breakdown exposes the stark social and ecological ineptitudes of our dominant economic logics and compels reconsideration of them.

Relatedly, climate change's temporal urgency and massive spatial scale can disable our sense of individual efficacy. One of my student's comments epitomised this: 'on a daily basis, I feel like I'm not doing enough, I'm not achieving enough to create this huge impact I'm waiting for. Though I feel like I'm not doing enough, I don't know what else I can do.' Neoliberal societies' emphasis on individual responsibility for this systemic problem burdens people with guilt, as this statement from my class indicates: 'I questioned whether my actions were creating any positive change. Was I simply another facet at the root of this problem?' As another commented, the incapacity of individuals to prevent rapid planetary breakdown leaves people tormented: 'climate change is huge, overwhelming, and I feel frustrated and angry. I am extremely cynical that humanity will do anything to mitigate climate change before it is too late.'

Feeling overwhelmed is one of the most frequent emotions I hear people express in response to climate change, and it is an apt word given it originally meant being inundated by water. In the context of melting glaciers, rising sea levels, bigger storm surges and more intense flooding, being overwhelmed by climate change will be a literal experience for far too many. For those of us learning about such events, whether through the media, in classrooms or otherwise, climate change can feel like drowning, flailing as we are engulfed with distressing information and immersed in political inertia. For example, climate scientist John Fasullo feels 'a flood of, what are at times, contradictory emotions.'[28] As cultures of individualism intersect with climate change's global and epochal scales, feelings of frustration, powerlessness, and hopelessness can submerge our sense of self.

This overwhelming affectivity of climate change can force us to reconsider our place within the world, and this sense of the shrinking or sinking of the self can operate on our embodied imaginaries. I myself have experienced this, noting in a journal one day:

> Sometimes when I think of climate change, I see this dark, vague, tsunami towering behind me, a frothing wall of utter destruction of which we have felt tremors, but by turning our backs, have not fully comprehended. I catch glimpses of it over my shoulder, about to crash down upon me, obliterating everything, but in front of me, life goes about its daily flow, oblivious to the imminent disaster.

As I (re)write this paragraph in 2020, the news tells me that the Greenland ice sheet lost *one million tonnes of ice per minute* throughout 2019,[29] and I am reminded of one of my student's comments: 'I feel small in the face of climate change.'

However, as the overwhelming affective agency of climate change can dissolve our individuality, it can bring our attention to otherwise over-looked relations. Grief is an emotional response to loss, and loss occurs 'when people are dispossessed of things that they value, and for which there are no commensurable substitutes.'[30] Climate change can threaten all elements of our lives, including 'phenomena that constitute the meaning of entire societies,' such as cultures, social cohesion and identities.[31] When we appreciate that climate is a set of relations and climate change a disruption of those relations, it becomes apparent that climate change *is* loss.[32] As such, experiences of climate grief can involve 'diagnosing' relationships: articulating or recognising relationships at the same moment as realising that they are under threat or ailing.[33] Consider this student's end-of-semester reflection, which expresses how climate change can lead to the loss of entire life-worlds:

It struck me that the world as we know it, and the way we live, is going to dramatically change. Even if we stop our current globally destructive practices right now, the climatic flow on effects will continue into the future.

Such experiences of anticipatory grief – the recognition that massive losses have already been set in motion – are a common experience when encountering climate change. Yet, while painful, climate grief can be 'a path to understanding entangled shared living and dying,'[34] as another student's comment articulates: 'I feel emotionally tied to the fate of our environment.' For these reasons, Ashley Cunsolo argues that we need to engage in the work of mourning climate change, and that engaging with our climate grief can be a promising political process:

> Re-casting climate change as the work of mourning means that we can share our losses, and encounter them as opportunities for productive and important work...In mourning [we] lose our former selves...We are changed internally and externally by the loss in ways that we cannot predict or control, and in ways that may be disorienting, surprising, or completely unexpected. These responses [can leave] us more open to...our transcorporeal connections with all bodies, species, and life forms.[35]

As Cunsolo articulates so eloquently, our encounters with the devastation of climate change can contribute to undermining individualism and anthropocentrism, and in this way, climate grief can contribute to the decomposition of discrete selves.

These experiences of loss not only generate more ecological ways of relating but can also contribute to the articulation of collective climate capacities. The distress generated through confronting our individual inefficacy can enable explorations of collectively distributed responsibility. Consider the following comment from a class discussion about the ways climate denial is socially constituted:

> the concept of a spectrum of denial is important because it breaks it down from good and evil, black and white; there's so much more to denial. We're all guilty of being deniers. Even though we get it, there's times life intervenes and we forget, the system twists our arms, forces us, to go against our conscious choice.

In these ways, our worries about the future and our incapacities to intervene in any meaningful way can generate critical analyses of capitalism and the articulation of more diffused notions of responsibility. For example, in the face of the most overwhelming of situations, school strikers insist that 'like the oceans, we rise.' Such emotionally charged social movements demonstrate that while climate change's immersive anguish can dissolve our sense of individual capacity, this disorientation can enable more collective, climate-capable, subjectivities to surface[36] (which I explore more in Chapter 5).

***

As unnerving and debilitating as the overwhelming experiences of frustration, guilt, grief, and anxiety explored above are, not all encounters with climate change are this straightforward. Climate change can be even more confusing, bewildering, subtle and/or elusive. We do not always notice how climate change is affecting us, nor are we always able to name such experiences, codify them as distinct emotions or convincingly analyse how they are operating or what they are doing to us. In this sense, encountering climate change not only disturbs our affective regimes, it undermines our self-reflective and analytical capacities. Climate change, following Naomi Klein, 'changes everything.'[37] The disruptions of everything we know and take for granted make climate change far more than what our existing – and even future – understandings of emotions, psychology and culture can accurately and consistently theorise. We are shaping Earth into an unrecognisable, and potentially uninhabitable, planet; no theory, no stories, no lived experience can fully account for these geological disorders. Climate change is excessive, boundless and dizzying, and human experiences of this can be unfathomable, indescribable and mysterious.

This is true for those experiencing climate change in its myriad manifestations, as well as for those like myself who seek to study and understand such experiences. One moment from our class speaks to the limits of human efforts to comprehensively know climate change. Following a lecture that included a distressing image of a family sheltering from the Dunalley bushfires (in Tasmania) in a dam in 2013,[38] one student mentioned in our tutorial that he was 'in a fiery mood.' It was early October (mid-spring), and the day before our class a planned fire in Lancefield (just 70 kilometres from Melbourne) exceeded authorities' control and went on to burn over 4,000 hectares, prompting the official start of the bushfire season to be moved forward a month.

What was going on? Potentially these events permeated him through media such as the photograph in the lecture and news reports of the

Lancefield fires; yet these events were geographically, and even temporally, distant (unlike the experiences of the 2019/2020 bushfires discussed in Chapter 2, where the smoke swirled all around and through people). While bushfires might also generate resigned, panicked or other moods, the use of 'fiery' to describe a mood suggests that the aggravated and unpredictable intensities of combustion can exceed the spaces, times and bodies usually considered within their bounds. This apparently momentary encounter with climate change could therefore be understood as a process which extended over a period of years and across the mountain range that separates Melbourne and Lancefield, or the sea the separates Tasmania from mainland Australia. This transferal and reconfiguration of intensity across kinds illustrates the transtemporal, transpatial and transcorporeal nature of climate change. It further indicates the ambiguous distinction between meteorological conditions and affective forces, as the smoke, haze, charcoal and cinders of the fires potentially infiltrated this student and inflamed his temper. But of course, this is somewhat speculative. Perhaps it was purely coincidental, maybe his choice of 'fiery' to describe his mood was unrelated to the flames that had glimmered in our classroom; I can't be sure. Climate change exceeds human capacities to neatly and comprehensively trace, document, describe and analyse.

Another incident from our class has stuck with me for years because I struggle to make sense of it. In one of my tutorials there were a lot of student absences in the first few weeks of semester, but it was not the same students missing class each week. Some would show up one week, then not again for a while, then suddenly arrive energised and passionate, resulting in each week being composed of a unique class cohort. At the start of class one afternoon a student referred to this in an off-hand comment: 'this class is a bit strange, people coming in and out. I guess that's a bit like climate change, you know, a bit vague, so many aspects, so much information, hard to grasp.' This comment has, for many years, troubled my capacity to interpret its significance. It had never aligned with the analysis discussed above, where particular features of the climate change problem generate relatively identifiable, and fairly logical, emotional responses. Was he saying that his experience of the class felt like climate change? Or was he saying that students were not attending because the course content is so challenging? Was he speculating that the inconsistent movement of students echoed planetary atmospheric disruptions? Or was it all of these things? After struggling for years to interpret, to categorise, to tame, this moment, I realised that in some ways I cannot do this, but at the same time, it is at least partly self-evident. He hits the nail on the head: climate change is hard to grasp. It is elusive, dispersed, slippery, vague, strange, overwhelming. And in this sense, the moment perhaps articulates most clearly how climate change manifests not just as internalised painful emotions, but as affective atmospheres: diffuse, nebulous, enigmatic, ambiguous, enveloping forces that enable, reconfigure and emerge from embodied relationships.[39] The affective disturbances that encounters with climate change generate arise from our planetary transcorporeality, and further entangle us with climate change. Given the unknowable complexity of the climate, we cannot always fully understand the ways we are affected by climate change.

***

Sara Ahmed reminds us that the etymology of 'emotion' is to move outwards.[40] In the conversations I have with people, they frequently describe climate change as incomprehensible: emotionally, morally and cognitively. Accompanying our discussions about such disconcertment, people sigh; smile; sweat; frown; pause; laugh; cry; lean back in their chairs; wriggle in their chairs; close their eyes; rub their eyes; roll their eyes; wipe tears from their eyes; establish, maintain or avert eye contact; hug each other; turn away from each other; listen or talk over each other; get up and leave; put their head in their hands, or on the table; stare at the ceiling; shrug their shoulders; slump their shoulders.

Climate change moves us. Even in 'climate-controlled' spaces we can encounter climate change in emotional and embodied ways. These affective atmospheres of climate change, which we continually co-re-create, both condense in on us and move us outwards. Encountering climate change can enrol us in metamorphic practices through which our sense of self variously shrinks, dissolves, expands and/or diffuses. This existential shapeshifting can disable our individualised sense of self and personal agency. Climate change's temporal urgency and global scale, its inherent uncertainty, its nature as a collective action problem and its relational composition can rupture neoliberal worlds that promise privileged people individual success and bright futures. These encounters can trouble settler-colonial and modern-industrial myths of certainty, progress and entitlement. As such, climate change can enroll us in processes that Rosi Braidotti terms *dis-identification*: the breakdown of established ways of being, identifying and relating.[41]

The destabilisation of such anthropocentric worldviews can enable us to identify and appreciate relations with others and provide space for cultivating more collective and ecological ways of being. Climate anxiety, if carefully engaged with, can be a catalyst for e-motional movements for climate justice, such as the youth climate strikes, which connect people and places around the world through shared experiences of planetary panic. Thus, despite the devastation that climate change generates, rather than seeking to heal or treat climate anxiety, we must see it as a promising opportunity to intervene in the very conditions that lead to climate change: illusions of mastery, desires for control and broader logics of extractivism.

Climate change counters us; through our affective engagements with climate change, we become climate-changed. Rather than climate change engagement consisting of humans containing or domesticating the climate, these emotional and affective responses speak to climate change's architectural capacities to tame, shape and direct us in ways we cannot fully understand, anticipate or imagine. By attuning to the uncanny, disorienting and devastating experiences of encountering climate anxiety, we might begin 'the work of mourning'[42] necessary for learning to live with climate change.

Notes

1 Lauren Rickards and Elspeth Oppermann, "Battling the Tropics to Settle a Nation: Negotiating Multiple Energies, Frontiers and Feedback Loops in Australia," *Energy Research & Social Science* 41 (2018): 100, https://doi.org/10.1016/j.erss.2018.04.038.

2 Blanche Verlie, "'Climatic-Affective Atmospheres': A Conceptual Tool for Affective Scholarship in a Changing Climate," *Emotion, Space and Society* 33 (2019), https://doi.org/10.1016/j.emospa.2019.100623.

3 David Rousell, Amy Cutter-Mackenzie, and Jasmyne Foster, "Children of an Earth to Come: Speculative Fiction, Geophilosophy and Climate Change Education Research," *Educational Studies* 53, no. 6 (2017): 666, https://doi.org/10.1080/00131946.2017.1369086.

4 Stephen Healy, "Atmospheres of Consumption: Shopping as Involuntary Vulnerability," *Emotion, Space and Society* 10, Supplement C (2014): 37, https://doi.org/10.1016/j.emospa.2012.10.003.

5 Lori Gruen, "Expressing Entangled Empathy: A Reply," *Hypatia* 32, no. 2 (2017).

6 Nino Antadze, "Who Is the Other in the Age of the Anthropocene? Introducing the Unknown Other in Climate Justice Discourse," *The Anthropocene Review* 6, no. 1–2 (2019), https://doi.org/10.1177/2053019619843679.

7 Petra Tschakert et al., "One Thousand Ways to Experience Loss: A Systematic Analysis of Climate-Related Intangible Harm from around the World," *Global Environmental Change* 55 (2019), https://doi.org/10.1016/j.gloenvcha.2018.11.006.

8 Lesley Head, *Hope and Grief in the Anthropocene: Re-Conceptualising Human–Nature Relations* (New York and Milton Park: Taylor & Francis, 2016).

9 Julietta Singh, *Unthinking Mastery: Dehumanism and Decolonial Entanglements* (Durham: Duke University Press, 2018); Val Plumwood, *Feminism and the Mastery of Nature* (London: Routledge, 1993).

10 Andrei Israel and Carolyn Sachs, "A Climate for Feminist Intervention: Feminist Science Studies and Climate Change," in *Research, Action and Policy: Addressing the Gendered Impacts of Climate Change*, ed. Margaret Alston and Kerri Whittenbury (Dordrecht, Heidelberg, New York and London: Springer, 2013).

11 Panu Pihkala, *Climate Anxiety* (Helsinki: MIELI Mental Health Finland, 2019); Elin Kelsey, "Propagating Collective Hope in the Midst of Environmental Doom and Gloom," *Canadian Journal of Environmental Education (CJEE)* 21 (2017).

12 Maan Barua, "Encounter," *Environmental Humanities* 7 (2015); Donna Haraway, *When Species Meet* (Minneapolis: University of Minnesota Press, 2008).

13 Panu Pihkala, "Anxiety and the Ecological Crisis: An Analysis of Eco-Anxiety and Climate Anxiety," *Sustainability* 12, no. 19 (2020).

14 Bas Verplanken and Deborah Roy, "'My Worries Are Rational, Climate Change Is Not': Habitual Ecological Worrying is an Adaptive Response," *PLOS ONE* 8, no. 9 (2013), https://doi.org/10.1371/journal.pone.0074708; Pihkala, "Anxiety and the Ecological Crisis: An Analysis of Eco-Anxiety and Climate Anxiety."

15 Rosemary Randall, "Climate Anxiety or Climate Distress? Coping with the Pain of the Climate Emergency," (October 19 2019). https://rorandall.org/2019/10/19/climate-anxiety-or-climate-distress-coping-with-the-pain-of-the-climate-emergency/.

16 Birgitte Schepelern Johansen, "Locating Hatred: On the Materiality of Emotions," *Emotion, Space and Society* 16 (2015): 50, https://doi.org/10.1016/j.emospa.2015.07.002.

17 IPCC, *Climate Change 2014: Synthesis Report. Contribution of Working Groups I, II and III to the Fifth Assessment Report of the Intergovernmental Panel on Climate Change*, Intergovernmental Panel on Climate Change (Geneva, 2014), 48 & 60.

18 Josephine Chu, "'Eco-Anxiety' over Climate Change Causing Stress, Panic, Experts Say," *Medill News Service* (August 28 2019). https://dc.medill.northwestern.edu/blog/2019/08/28/eco-anxiety-over-climate-change-causing-stress-panic-experts-say/#sthash.Szkpk24B.Qat17vF6.dpbs; Marianne PRT, "I Have Eco-Anxiety," *Voices of Youth* (April 2 2019). https://www.voicesofyouth.org/blog/i-have-eco-anxiety.

19 Astrida Neimanis, *Bodies of Water: Posthuman Feminist Phenomenology* (London: Bloomsbury Academic, 2017).

20 Eve Tuck, Marcia McKenzie, and Kate McCoy, "Land Education: Indigenous, Post-Colonial, and Decolonizing Perspectives on Place and Environmental Education Research," *Environmental Education Research* 20, no. 1 (2014), https://doi.org/10.1080/13504622.2013.877708.

21 Margolin, Jamie. *Jamie Margolin's 2019 Congressional Testimony*, Voices Leading the Next Generation on the Global Climate Crisis, United States House of Representatives Committee on Foreign Affairs (2019).

22 Robert Brulle and Kari Marie Norgaard, "Avoiding Cultural Trauma: Climate Change and Social Inertia," *Environmental Politics* 28, no. 5 (2019), https://doi.org/10.1080/09644016.2018.1562138.

23 Head, *Hope and Grief in the Anthropocene: Re-Conceptualising Human–Nature Relations.*

24 Greta Thunberg, *No One Is Too Small to Make a Difference* (Penguin, 2019), 58.

25 Kyle Whyte, "Indigenous Science (Fiction) for the Anthropocene: Ancestral Dystopias and Fantasies of Climate Change Crises," *Environment and Planning E: Nature and Space* 1, no. 1–2 (2018), https://doi.org/10.1177/2514848618777621; Heather Davis and Zoe Todd, "On the Importance of a Date, or Decolonizing the Anthropocene," *ACME: An International E-Journal for Critical Geographies* 16, no. 4 (2017).

26 Davis and Todd, "On the Importance of a Date, or Decolonizing the Anthropocene."

27 Thunberg, *No One Is Too Small to Make a Difference*, 58. There is an irony here which is that the old adage is right, the sky *is* the limit, but we are much more closely entangled with the sky than imagined.

28 John Fasullo, "ITHYF5," *Is This How You Feel?* (2020). https://www.isthishowyoufeel.com/ithyf5.html.

29 Damian Carrington, "Greenland Ice Sheet Lost a Record 1m Tonnes of Ice Per Minute in 2019," *The Guardian*, August 21, 2020, https://www.theguardian.com/environment/2020/aug/20/greenland-ice-sheet-lost-a-record-1m-tonnes-of-ice-per-minute-in-2019; Ingo Sasgen et al., "Return to Rapid Ice Loss in Greenland and Record Loss in 2019 Detected by the Grace-Fo Satellites," *Communications Earth & Environment* 1, no. 1 (2020), https://doi.org/10.1038/s43247-020-0010-1.

30 Jon Barnett et al., "A Science of Loss," *Nature Climate Change* 6, no. 11 (2016): 977.

31 Barnett et al., "A Science of Loss," 977.

32 Ashlee Cunsolo Willox, "Climate Change as the Work of Mourning," *Ethics & the Environment* 17, no. 2 (2012), https://doi.org/10.2979/ethicsenviro.17.2.137.

33 Susan Ruddick, "Rethinking the Subject, Reimagining Worlds," *Dialogues in Human Geography* 7, no. 2 (2017), https://doi.org/10.1177/2043820617717847.

34 Donna Haraway, *Staying with the Trouble: Making Kin in the Chthulucene* (Durham and London: Duke University Press, 2016), 39.

35 Cunsolo Willox, "Climate Change as the Work of Mourning," 145, 57.

36 Sara Ahmed, "Collective Feelings: Or, the Impressions Left by Others," *Theory, Culture & Society* 21, no. 2 (2004), https://doi.org/10.1177/0263276404042133.

37 Naomi Klein, *This Changes Everything: Capitalism Vs. The Climate* (New York: Simon & Schuster, 2014).

38 Jon Henley, "Guardian Firestorm Film About Tasmanian Bushfire in Competition," *The Guardian*, June 5, 2013. https://www.theguardian.com/

film/2013/jun/05/guardian-firestorm-tasmania-dunalley-competition-docfest.
39 Ben Anderson, "Affective Atmospheres," *Emotion, Space and Society* 2, no. 2 (2009), https://doi.org/10.1016/j.emospa.2009.08.005; Verlie, "'Climatic-Affective Atmospheres': A Conceptual Tool for Affective Scholarship in a Changing Climate."
40 Ahmed, "Collective Feelings: Or, the Impressions Left by Others."
41 Rosi Braidotti, *The Posthuman* (Cambridge and Malden: Polity Press, 2013).
42 Cunsolo Willox, "Climate Change as the Work of Mourning."

References

Ahmed, Sara. "Collective Feelings: Or, the Impressions Left by Others." *Theory, Culture & Society* 21, no. 2 (2004): 25–42. https://doi.org/10.1177/02 63276404042133.

Anderson, Ben. "Affective Atmospheres." *Emotion, Space and Society* 2, no. 2 (2009): 77–81. https://doi.org/10.1016/j.emospa.2009.08.005.

Antadze, Nino. "Who Is the Other in the Age of the Anthropocene? Introducing the Unknown Other in Climate Justice Discourse." *The Anthropocene Review* 6, no. 1–2 (2019): 38–54. https://doi.org/10.1177/2053 019619843679.

Barnett, Jon, Petra Tschakert, Lesley Head, and W. Neil Adger. "A Science of Loss." *Nature Climate Change* 6, no. 11 (2016): 976–78.

Barua, Maan. "Encounter." *Environmental Humanities* 7 (2015): 265–70.

Braidotti, Rosi. *The Posthuman.* Cambridge and Malden: Polity Press, 2013.

Brulle, Robert, and Kari Marie Norgaard. "Avoiding Cultural Trauma: Climate Change and Social Inertia." *Environmental Politics* 28, no. 5 (2019): 886–908. https://doi.org/10.1080/09644016.2018.1562138.

Carrington, Damian. "Greenland Ice Sheet Lost a Record 1m Tonnes of Ice Per Minute in 2019." *The Guardian*, August 21, 2020. https://www.theguardian.com/environment/2020/aug/20/greenland-ice-sheet-lost-a-record-1m-tonnes-of-ice-per-minute-in-2019.

Chu, Josephine. "'Eco-Anxiety' over Climate Change Causing Stress, Panic, Experts Say." *Medill News Service*, August 28, 2019. https://dc.medill.northwestern.edu/blog/2019/08/28/eco-anxiety-over-climate-change-causing-stress-panic-experts-say/#sthash.Szkpk24B.Qat17vF6.dpbs.

Cunsolo Willox, Ashlee. "Climate Change as the Work of Mourning." *Ethics & the Environment* 17, no. 2 (2012): 137–64. https://doi.org/10.2979/ethicsenviro.17.2.137.

Davis, Heather, and Zoe Todd. "On the Importance of a Date, or Decolonizing the Anthropocene." *ACME: An International E-Journal for Critical Geographies* 16, no. 4 (2017): 761–80.

Fasullo, John. "ITHYF5." *Is This How You Feel?* (2020). https://www.isthishowyoufeel.com/ithyf5.html.

Gruen, Lori. "Expressing Entangled Empathy: A Reply." *Hypatia* 32, no. 2 (2017): 452–62.

Haraway, Donna. *Staying with the Trouble: Making Kin in the Chthulucene.* Durham and London: Duke University Press, 2016.

Haraway, Donna. *When Species Meet.* Minneapolis: University of Minnesota Press, 2008.

Head, Lesley. *Hope and Grief in the Anthropocene: Re-Conceptualising Human–Nature Relations.* New York and Milton Park: Taylor & Francis, 2016.

Healy, Stephen. "Atmospheres of Consumption: Shopping as Involuntary Vulnerability." *Emotion, Space and Society* 10, Supplement C (2014): 35–43. https://doi.org/10.1016/j.emospa.2012.10.003.

Henley, Jon. "Guardian Firestorm Film About Tasmanian Bushfire in Competition." *The Guardian*, June 5, 2013. https://www.theguardian.com/film/2013/jun/05/guardian-firestorm-tasmania-dunalley-competition-docfest.

IPCC. *Climate Change 2014: Synthesis Report. Contribution of Working Groups I, II and III to the Fifth Assessment Report of the Intergovernmental Panel on Climate Change.* Geneva: Intergovernmental Panel on Climate Change, 2014.

Israel, Andrei, and Carolyn Sachs. "A Climate for Feminist Intervention: Feminist Science Studies and Climate Change." In *Research, Action and Policy: Addressing the Gendered Impacts of Climate Change*, edited by Margaret Alston and Kerri Whittenbury, 33–51. Dordrecht, Heidelberg, New York and London: Springer, 2013.

Johansen, Birgitte Schepelern. "Locating Hatred: On the Materiality of Emotions." *Emotion, Space and Society* 16 (2015): 48–55. https://doi.org/10.1016/j.emospa.2015.07.002.

Kelsey, Elin. "Propagating Collective Hope in the Midst of Environmental Doom and Gloom." *Canadian Journal of Environmental Education (CJEE)* 21 (2017): 23–40.

Klein, Naomi. *This Changes Everything: Capitalism Vs. The Climate.* New York: Simon & Schuster, 2014.

Margolin, Jamie. *Jamie Margolin's 2019 Congressional Testimony.* Voices Leading the Next Generation on the Global Climate Crisis. United States House of Representatives Committee on Foreign Affairs, 2019.

Neimanis, Astrida. *Bodies of Water: Posthuman Feminist Phenomenology.* London: Bloomsbury Academic, 2017.

Pihkala, Panu. "Anxiety and the Ecological Crisis: An Analysis of Eco-Anxiety and Climate Anxiety." *Sustainability* 12, no. 19 (2020): 7836.

Pihkala, Panu. *Climate Anxiety.* Helsinki: MIELI Mental Health Finland, 2019.

Plumwood, Val. *Feminism and the Mastery of Nature.* London: Routledge, 1993.

PRT, Marianne. "I Have Eco-Anxiety." *Voices of Youth*, April 2, 2019. https://www.voicesofyouth.org/blog/i-have-eco-anxiety.

Randall, Rosemary. "Climate Anxiety or Climate Distress? Coping with the Pain of the Climate Emergency," October 19, 2019. https://rorandall.org/201 9/10/19/climate-anxiety-or-climate-distress-coping-with-the-pain-of-the-climate-emergency/.

Rickards, Lauren, and Elspeth Oppermann. "Battling the Tropics to Settle a Nation: Negotiating Multiple Energies, Frontiers and Feedback Loops in Australia." *Energy Research & Social Science* 41 (2018): 97–108. https:// doi.org/10.1016/j.erss.2018.04.038.

Rousell, David, Amy Cutter-Mackenzie, and Jasmyne Foster. "Children of an Earth to Come: Speculative Fiction, Geophilosophy and Climate Change Education Research." *Educational Studies* 53, no. 6 (2017): 654–69. https:// doi.org/10.1080/00131946.2017.1369086.

Ruddick, Susan. "Rethinking the Subject, Reimagining Worlds." *Dialogues in Human Geography* 7, no. 2 (2017): 119–39. https://doi.org/10.1177/204382 0617717847.

Sasgen, Ingo, Bert Wouters, Alex Gardner, Michalea King, Marco Tedesco, Felix Landerer, Christoph Dahle, Himanshu Save, and Xavier Fettweis. "Return to Rapid Ice Loss in Greenland and Record Loss in 2019 Detected by the Grace-Fo Satellites." *Communications Earth & Environment* 1, no. 1 (2020): 8. https://doi.org/10.1038/s43247-020-0010-1.

Singh, Julietta. *Unthinking Mastery: Dehumanism and Decolonial Entanglements*. Durham: Duke University Press, 2018.

Thunberg, Greta. *No One Is Too Small to Make a Difference*. Penguin, 2019.

Tschakert, Petra, N.R. Ellis, C. Anderson, A. Kelly, and J. Obeng. "One Thousand Ways to Experience Loss: A Systematic Analysis of Climate-Related Intangible Harm from around the World." *Global Environmental Change* 55 (2019): 58–72. https://doi.org/10.1016/j.gloenvcha.2018.11.006.

Tuck, Eve, Marcia McKenzie, and Kate McCoy. "Land Education: Indigenous, Post-Colonial, and Decolonizing Perspectives on Place and Environmental Education Research." *Environmental Education Research* 20, no. 1 (2014): 1–23. https://doi.org/10.1080/13504622.2013.877708.

Verlie, Blanche. "'Climatic-Affective Atmospheres': A Conceptual Tool for Affective Scholarship in a Changing Climate." *Emotion, Space and Society* 33 (2019): 100623. https://doi.org/10.1016/j.emospa.2019.100623.

Verplanken, Bas, and Deborah Roy. "'My Worries Are Rational, Climate Change Is Not': Habitual Ecological Worrying is an Adaptive Response." *PLOS ONE* 8, no. 9 (2013): e74708. https://doi.org/10.1371/ journal.pone.0074708.

Whyte, Kyle. "Indigenous Science (Fiction) for the Anthropocene: Ancestral Dystopias and Fantasies of Climate Change Crises." *Environment and Planning E: Nature and Space* 1, no. 1–2 (2018): 224–42. https://doi.org/1 0.1177/2514848618777621.

4 Witnessing multiple climate realities

When first starting this course I thought of climate change in terms of facts and numbers. In my head, there was a climate problem. There was too much CO_2 in the atmosphere and humanity needed to stop emitting it. It was a scientific problem that could be fixed through a scientific response. But the complexity of the scientific information provided in graphs and figures was quite confusing and without detailed explanation, hard to understand. I left feeling very unsettled and the feelings of uncertainty and confusion explained to me why so many people choose to ignore climate change. I think the bleak future predictions and cynical framing of climate change are some of the reasons for society's denial and lack of behavioural change.

I am constantly butting heads with sceptics and non-believers (particularly my father-in-law) regarding climate change. It is so frustrating that fellow inhabitants don't understand the magnitude of the situation, and worse still, they don't care to learn more about.

I find climate inaction so annoying and perplexing. Even though people know about climate change and express concern, they still don't take action – why? This has bugged me for some time.

Throughout the semester, I have struggled to understand how those who are less vulnerable and more resilient for the most part seem disenchanted with the idea of climate change, even when it is occurring before their eyes. I find this really fascinating, that there is a problem, but just because it isn't immediate or is associated with some 'uncertainty' it is dismissed.

This is what I have learned, that the more I know, the more I don't know and that's a scary and depressing thought.

DOI: 10.4324/9780367441265-4

I feel that Australian society has buried their heads in the sand at the thought of climate change and this has to do with certain Australian leaders not valuing a sustainable future. Our national identity is somewhat influenced by the government of the day, whose policies are creating confusion even though people believe climate change is an issue.

To some extent every one of us is a denier and that fact indicates the overarching problem of climate change itself. A lack of accountability on a global scale.

This hit me a little hard.

The problem is so immense and entwined via almost any issue you can think of in this world.

There is not enough space to talk about how climate change makes you feel, and I think that's important. If scepticism is embedded in our national identity, we almost need climate change therapy.

I've realised learning about climate change is a traumatic process. I was thinking of the dark, foreboding nature of climate change, its creeping horror masked by invisibility in the here-and-now of hyperconsumptive capitalism. Sometimes I see climate change as a chasm opening up before me, and I stand on a precipice overlooking the deep ravine, teetering on the edge.

Ignorance is bliss, because this semester has been a roller coaster of emotions. I have been overwhelmed by joy, fear, and passion. The tutorials were scary; though I feel I learnt so much from simply listening to people speak from their heart and experience always leaving feeling refreshed.

Coming into this course with a completely different background to all the other people in the class and not knowing anyone was actually quite daunting. But after listening to everyone talk and interact with each other in the classes, I soon found that everyone was on the same page and this feeling of being overwhelmed diminished somewhat.

I have never felt so emotionally connected to a concept I have learnt in a classroom. Climate is not just the weather you check on an app so that you can dress accordingly. Climate is a live or die reality, unpredictable and raw. But I easily forget the scope of our environment and its presence with me. And if I as a student who discusses and thinks about this subject four or more days a week can forget, then it's no wonder the rest of the world does too.

I have learned that this social response is hard to explain, humans are complicated organisms and there are many different theories to explain why someone is in climate denial, I just hope it is not because they have stopped caring about the environment and future generations.

***

'Reality' and 'truth' have high stakes when it comes to climate change. Al Gore's international NGO is called the Climate Reality Project, and Extinction Rebellion's first demand is 'tell the truth.' The truth and reality such efforts argue for is that climate change is real, is caused by things (some) humans do, is already happening, and demands immediate and serious response. Such ardent emphasis on accepting reality speaks to the extent to which climate advocacy has had to position itself against highly funded, systemic and institutionalised climate denial. For decades, activists have argued that we need everyone, especially politicians, to 'unite behind the science' (to use Greta Thunberg's words). However, in these efforts to counteract denial, we can slip into implying that science is the sole arbiter of 'the' climate reality and 'the' climate truth.

Lived experiences of climate change demonstrate that everyday climate realities are plural, diverse, nuanced and unique, and they far exceed what climate science discloses.[1] For example, for some women the reality of 'recovering' from the 2009 Black Saturday bushfires in Australia included the experience of new or increased domestic violence.[2] For some people in Puerto Rico living in the aftermath of Hurricanes Irma and Maria, their climate realities can include overcrowded, dirty and damaged housing, leading to the loss of privacy, pride, comfort and free time, all of which disturb interpersonal relationships and identities.[3] These lived realities are not incompatible with the scientific reality; but they are far more specific, inhabited, emplaced and inter/personal than what science's models and graphs convey. One of the reasons that climate science has not sufficiently engaged people is because it cannot fully account for such lived realities of climate change, and it thus continues to feel irrelevant, distanced and abstract.

In the last chapter I argued that climate change will always be more than we can understand and that letting go of the desire to be able to fully know, predict and thus control the climate is an important ethical practice. However, this does not mean we should dismiss the knowledges we do and can have. Seeking, crafting, enacting and sharing climatic knowledges are evidently crucial practices in the catastrophic times in which we live. Indeed, as one of my students put it, 'our knowledges and ignorances about climate change will impact who will live and who will die.' Although there are multiple and persistent climate lies, ignorances and denials, this does not mean there is a singular truth of a universal climate reality. This chapter explores additional modes of coming to know climate change and how such different knowledges participate in the realisation of multifaceted, vivid, compelling climate realities.

I discuss witnessing as a practice that can enable us to acknowledge, articulate, inhabit, enact and respond to these lively, complex, intersecting climate worlds. Witnessing authenticates, validates or legitimates particular experiences and/or accounts of reality, and is thus how truths and knowledges are established, maintained and normalised.[4] But witnessing involves more than absorbing and recalling factual information; witnessing is an affective labour that requires response. We *bear* witness; to be a witness is to be more than just a bystander. Witnessing climate change enables a strong rejection of climate denial, but in order to do so, it requires facing up to the violences, traumas and injustices of climate change, sitting with the discomfort of this, and crafting collaborative and respectful responses. As such, witnessing climate change draws us into ethical relationships with other humans and the more-than-human world. We stand *with* others – friends, students, peers, future generations, those on the front lines, other species, ecosystems – as we witness the suffering extractive cultures inflict upon them.[5] 'Wit(h)nessing'[6] climate change therefore requires listening to others' experiences and validating the diverse climate realities they inhabit. It moves beyond exclusionary notions of a singular and universal climate truth, and as such is a move from consumptive and replicative climate knowledge to collaborative, enactive and regenerative climate knowledges.

Witnessing usually involves the acknowledgement and/or production of testimony, and in anthropocentric accounts, both the provider of testimony and the witness are often assumed to be individuated humans. Yet witnessing climate change requires attuning to all kinds of testimony and all kinds of witnesses. We might find testimony and/or witnesses of climate change in the rings of trees, in ice core samples, in bone dry riverbeds, in saline drinking water, in mass animal deaths, in the changing gender of sea turtles, in bleached coral reefs, in collapsing roads. This can lead to cascades of more-than-human witnessing and testimony: statistics, graphs, media, discussions, art, activism, curricula, policy, migrating animals, disappearing ecosystems, etc. The magpie caught on camera mimicking fire truck sirens during Australia's 2019/2020 bushfires demonstrates this: human responses to the bushfires sent multitudes of fire trucks; the magpie perceived this change in their local soundscape (and probably noticed the fires) and decided to mimic it; a local man noticed this novel and unnerving event, filmed it and posted it to social media; other humans viewed the magpie witnessing the fire trucks, and in turn wrote news stories so that it could be witnessed by others.[7]

In this chapter I explore multiple practices of witnessing climate change, including through affective, ethical, subconscious and spirited modes of knowing. I demonstrate that rather than observing climate from afar, witnessing is enacted through being part of climate change. It is an affective labour that implicates us in climate change, enrols us in relations with others, and changes who we are. I argue that if we conceptualise climate science not as a dispassionate practice of *knowing about* climate change, but of *witnessing* it, we may be better placed to integrate diverse climate knowledges and support people to engage in the emotional work of learning to live with climate change.

***

Climate science is supposed to be objective, which in turn is supposed to mean value free, dispassionate, and conducted by human scientists from a position of externality to the climate. The aggressive undermining of climate science by institutionalised climate denial has led to advocates and scientists seeking to bolster this reputation of climate science. As such, the discourses of climate science enact, and indeed epitomise, what Donna Haraway calls 'the God's eye view,' where cognitive mastery can be achieved through observation from a distance.[8] This is a gaze constructed through extractivist-technocentrism that asserts and normalises supposedly disembodied and global perspectives, perspectives which are in fact distinctly masculine, Western and managerial. A friend of mine, Nathan Eizenberg, an atmospheric scientist who works with climate models, once commented on the ways that climate science's view from 'everywhere and nowhere'[9] can overshadow our embodied engagements with climate:

Climate modelling inverts your perspective on reality. Sometimes I can't help but think about equations and model parametrisations when I am hiking on the razorback of some national park trail and I feel the rush of a leeward wind pass through my hair...so sad.

The God's eye view is exemplified in diagrams of Earth's atmosphere that look down upon the planet from a celestial viewpoint, but this philosophy of knowing climate through disembodied distance is maintained throughout the whole endeavour of climate science.[10] Of course, such diagrams offer important explanatory power that can enable people to connect their lived experiences to global phenomena (such as Eizenberg quoted earlier), and climate science more broadly provides absolutely integral knowledge about how the climate of our only possible home is changing rapidly and becoming literally uninhabitable.

However, through these notions of objectivity-via-separation, climate science normalises a sense that the best way to understand climate change is through extracting yourself from it. The God's eye view of climate science functions by asserting and maintaining a clear and consistent boundary between the human climate knower and the climate that they seek to know. Climate science's globalised viewpoints can therefore lead to oversights that fail to attend to the way particular bodies, in particular places, experience and know climate change.[11] This can reinforce, rather than overcome, apathy in the face of climate change; support myths that humans can and should be masters of the climate; and it also generalises from the privileged experiences of scientists which can erase diverse and disadvantaged perspectives.

In these ways, adherence to dispassionate climate science as the only mode of knowing climate change can function as a kind of climate denial. Commitment to the statistical definition of climate as the average weather means that we can slip into believing that 'climate cannot be experienced directly through the senses.'[12] I have heard advocates of climate science scorn people who speak of their experiences of climate change by informing them that this was an experience not of climate change but of weather.[13] But if we cannot experience climate change, why care about it? In such devoted attachment to correct scientific knowledge, we can exclude, ignore and even deny other knowledges, including those of people who are living with and responding to the realities of climate change in their everyday lives.

Consider the following two reflective comments from students at the completion of our class, which emphasise how their own fervent adherence to dispassionate climate science operated as a mode of climate denial:

> Before taking this course, I had subconsciously begun to consider my science work as 'real work' and begun to devalue the social sciences. My thought process was pretty narrow-minded: 'We don't have time for this! Climate change is happening right now! Let's do the science and get this thing solved!' Of course, we have 'done the science' and are continuing to do the science…I may have been using the unemotional, disinterested framework of science as a way of shielding myself from the sheer terror of climate change.

> At the start of this course, I thought we would be learning about better communication of science. Rather, we need to think harder about how and what we communicate. How can we connect on that emotional level? How do we get people to understand that we are scared too, and that they are not alone? In fact, my initial understanding – that science was the way to communicate about climate change – was a form of socio-psychological denial.

These two students' conclusions arose after a semester of collectively witnessing climate change in multiple ways. This included these students witnessing their own and others' affective unsettlement in our classes, such as those experiences discussed in Chapter 3 and which I discuss further below. These were experiences that dispassionate climate science was unable to fully anticipate, acknowledge or respond to.

Our modes of knowing climate change are always practices of affectively and transcorporeally entangling with it, and this includes scientific knowledges. We cannot ever extract ourselves from climate. Despite climate science's efforts at detachment, climate scientists themselves experience a range of distressing emotional experiences due to working everyday with the numbers and statistics describing our changing climate. Caught between the ideal of rational and dispassionate science and the world-shattering data they work with, climate scientists engage in considerable emotional labour in order to 'keep the heart a long way from the brain,' a kind of everyday emotional climate denial that enables scientists to keep going with the terrifying work of accurately calculating climate collapse.[14] In spite of such strategies, climate scientists are hyper-aware of the traumatic implications of climate change for the world. They are also increasingly seeing 'their own activities and choices captured in the curves of rising temperatures,' which is leading some of them to speak out politically and to reduce their personal and professional carbon footprints.[15] Further, climate change research – its physical infrastructures, processes and people – are increasingly being impacted by climate change.[16] In these ways, the actual humans who study climate are deeply emotionally, physically, intellectually and politically implicated in and affected by the climate, despite their efforts to methodologically distance themselves from it. All of this demonstrates that even the supposedly dispassionate practice of knowing climate change scientifically can be a form of witnessing climate change.

Conceptualising climate science as a practice of witnessing can ensure we better appreciate the affective energy required to engage in climate science, both by professional scientists and those of us engaging with their work, whether in classrooms or wider society. Climate science is a complex affective labour which tries to distance the knower in order to produce testimony of planetary trauma in the form of unemotional models, predictions, and statistics. But even the cold hard rationality of mathematics cannot contain climate change's pervasive affective agency. Climate change exceeds the efforts of science to dissociate from it. In multiple ways, climate change gets around, infiltrates and/or dissolves the scientist/climate distinction. To engage with climate science is therefore to engage in the transcorporeal emotional work of witnessing climate catastrophe and producing testimony of that. As such, creating opportunities for those engaging with climate science to have their affective experiences witnessed – acknowledged and validated – by others is important, because ignoring, denying or erasing them perpetuates the extractive violence of pretending we are not entangled with climate.

To acknowledge climate science as an embodied and emotional practice in these ways is not to deny its capacity to provide working, functioning, accurate knowledge. Unlike climate denial which refuses to know, witnessing climate change, including through scientific practice, is enacted through careful engagements with material worlds. Climate science scales up and consistently replicates one particular mode of this, and it is the consistent enactment of the God's eye view that enables science's powerful knowledges of how the planet's climate is changing.[17] However, this globalised worldview limits and denies other climate realities – including some of those its own methods generate, such as the existential anxiety of climate scientists. As Haraway articulates, 'an optics is a politics of positioning' and 'ways of seeing are ways of life.'[18] Thus, to witness more complex, multifaceted, lively climate realities, we have to re-position and/or blur the boundaries we enact when we seek to 'see,' i.e. know, climate change.[19]

***

Rather than looking *at* climate change, certain other-than-scientific knowledges can help us learn to see *through, with* or *as* climate change. Embodied empathy, moral reasoning, sociopolitical analyses and lived experience can contribute to alternative ways of being in relation with climate change. As such, they can enable us to better acknowledge and respond to our enmeshment within complex climate realities. For example, Nancy Tuana demonstrates how engaging with feminist philosophy and climate justice can enable us to 'see through the eye of Hurricane Katrina.'[20] Doing so reveals how racist urban planning policies can contribute to a Category Five hurricane flattening a city while sending inequality through the roof. As Tuana argues, Katrina clarifies that if we are to adequately respond to the ways planetary heating and systemic social marginalisation interact, we must refuse the society/environment binary and imagine and enact alternative climate–human relationships.

In this way, engaging climate change as a lens – or to be less ableist, as an apparatus for knowing the world – is a move away from the God's eye view which objectifies climate change. It is a form of witnessing where climate change, in all its complexity, becomes a kind of affective-ecological-ethical prosthetic that enables us to become differently entangled with climate change. Because we are a part – but not the entirety – of climate change, climate change is simultaneously what we seek to know as well as our means of knowing. Affective climate knowledges position us right in this mess, no longer looking on with the confidence of detachment but unavoidably embroiled in complicated ethical challenges. One of my student's comments depicts this witnessing-as-part-of-climate-change. Referencing her oscillating pessimism and hope, she ventured that:

> maybe we are just too close to see, but our input does matter and we can make a difference. Climate change can be the catalyst we need, a disaster response that fights inequality and builds a just economy through a common lens.

Echoing Tuana, the student articulates that climate change illuminates the stark hierarchies and violences that characterise the current global world order. Yet at the same time, she is so engaged with the issue that she loses her ability to ascertain whether efforts to intervene can be effective. This being 'too close to see,' an experience which arose through her community activism, environmental studies and climate anxiety, is an example of how affective climate knowledges emerge from and enrol us in different relationships and worlds. In this moment, the student enacts a 'partial perspective,'[21] an embodied and personally emplaced mode of witnessing which situates her as part of a proactive collective. Her comment articulates how the transcorporeal practice of witnessing climate change can only ever offer knowledges that obscure as well as reveal.[22]

Witnessing climate change is an active practice of knowing-through-relation that brings realities into being. As part of this, witnessing climate change acts on and changes the witness themself. The ways we 'reflect' on our own implication in climate denial is one example of this. This is powerfully articulated by another student's end of semester comment which drew on Kari Norgaard's scholarship:[23]

> reading Norgaard's chapter about climate denial resonated with me – perhaps studying environmental science, eating organic sourdough, occasionally using my 'keepcup' and riding my bike everywhere isn't enough? I could see myself in the characterisations she was depicting: climate change as background noise, avoiding thinking about it so as to not confront feelings of guilt and helplessness, concerned but apathetic. And yet, when she tells of public interest declining as scientific evidence mounts, I'm appalled. Like many of those in her ethnographic research, I'm able to live a 'double reality,' where knowledge of climate change is denied in favour of maintaining the comfortable, non-confrontational status quo.

This student goes on to state that exploring how he was living in two worlds 'challenged my personal identity and assumptions.' Witnessing his own inhabitation of these 'parallel universes'[24] forced him to contemplate his future, actions, identifications, what truths he believes and what worlds he wants to inhabit and create.

Similar to many of the statements in the vignettes, this student's comment demonstrates that witnessing climate change involves witnessing our own routine practices of denial. In so doing, it brings into being the multiple realities that we can live within: the 'climate-change-is-real' reality and the 'business-as-usual' reality. The experience of such conflicting climate realities is often described by people as 'like living in *The Matrix*,'[25] referencing the 1999 science fiction action film.[26] Such analogies assert that the 'real' reality of climate change is rendered invisible through complex systems of power and manufactured ignorance. For those who identify themselves as people who 'took the red pill,' *The Matrix* helps explain the challenges of being concerned about climate change yet living in a society of mass climate denial which 'doesn't bear any resemblance to our felt reality.'[27] In such discussions, often individual people are positioned as being either believers or deniers who live in one of these worlds, but not both. Yet, as the student cited earlier and Norgaard's research demonstrates, we can also inhabit both of these realities, explicitly acknowledging the truth of climate change in particular moments but living as if it is not happening most of the time.

As his comment indicates, engaging climate change as a lens through which to reflect on your own practices of denial also enacts a kind of 'diffractive' vision,[28] where, rather than running parallel, these two worlds collide and interfere with each other, producing overlapping, elusive, splintered and rippling realities. For journalist Brigid Delaney these two realities slammed into each other while she was on a boat crossing Sydney Harbour for a private wine tasting tour amidst the suffocating smoke of the 2019/2020 bushfires:

> We were all in our party dresses and chunky trainers, phones fully charged to maximise the Instagrammable location, only coughing a little bit although peoples' eyes were red and I noticed some fellow guests pulling on Ventolin inhalers...We stood near the pool, eating tiny food, drinking wine from large balloon glasses while ash flew from the sky, some of it landing in my drink... Every varietal had notes of bushfire...[Yet] influencers posed in the gloom on the jetty and by the swimming pool, seeing but refusing to see what was all around them: this red-raw sun, that dirty brown sky...Various people wandered up to us and said 'great day for it!' and 'beautiful weather' without irony. *How could they say that?*...We could have been excused for thinking we were crossing the Styx – the mystical Greek crossing into the Underworld.[29]

For Delaney, the floating flecks of forest dispersed a 'truth bomb' all around and indeed, all through, her: 'when wine turns to ash in your mouth, you can't deny the new reality anymore.' But Delaney's sense that they were leaving this world also emphasises that this reality can feel both irrefutable and decidedly *un*real.

To become aware that climate change is enveloping and even interpermeating us while the machinery of extractive capitalism continues unperturbed can feel surreal and alienating. Our practices of witnessing the 'climate-change-is-real' reality can merge into what we might term climate 'un/realities,' the realm of imagination, daydreams, and nightmares. For example, science communicator Joe Duggan has collated letters written by climate scientists responding to the question, 'how does climate change make you feel?' Stefan Rahmstorf, Head of Earth System Analysis at the Potsdam Institute for Climate Impact Research, responded by narrating a recurring dream:

> I'm going for a hike and discover a remote farm house on fire. Children are calling for help from the upper windows. So I call the fire brigade. But they don't come, because some mad person keeps telling them that it is a false alarm. The situation is getting more and more desperate, but I can't convince the firemen to get going. I cannot wake up from this nightmare.[30]

Australian climate scientist Joëlle Gergis recounted a similar experience in her piece *Witnessing the Unthinkable* which was written in the wake of the 2019/2020 bushfires and discussed the recently updated, and more terrifying, projections from climate science:

> I keep having dreams of being inundated. Huge, monstrous waves bearing down on me in slow motion…I watch as a colossal tsunami builds offshore. I panic, immediately sensing that I don't stand a chance. I watch the horizon disappear, before turning to bolt to higher ground. Around me, people are calmly going about their business.[31]

Relatedly, people sometimes tell me of surreal experiences when climate change has interrupted their everyday business-as-usual reality while they were awake. These are apocalyptic, melancholy and/or jarring scenes that might replay over and over, or erupt as if from nowhere, or perhaps be barely discernible, a niggling estrangement at the periphery of awareness. These are not daydreams so much as waking nightmares. Consider the following anecdote narrated in one of our classes which discusses the experience of studying climate change:

> It's like, on warm, sunny winter and early spring days, with the light glistening through young green leaves. Everyone is happy due to the nice weather. But knowing about climate change, you know

it means someone somewhere is not getting the rain they need. I felt the warm spring weather was just an indication that the fires will be worse this summer, and that if I looked hard enough, I felt I could see the trees burning rather than shimmering. So it's sort of, you can't enjoy it, it's an uneasiness amongst the glory that everyone else seems to be celebrating.

Some people tell me of the dystopian worlds they anticipate their as-yet-unconceived grandchildren living in. Others describe experiences of climate change enveloping, overshadowing or otherwise haunting them while they complete mundane everyday tasks like shopping or cleaning. In many of these cases, people are pre-emptively experiencing the affective intensities of possible futures. Yet chronic vivid nightmares of these kinds can also lead to actual climatic disasters feeling like déjà vu as they unfold, generating a strange, numb familiarity with unprecedented traumas.

Such experiences of climate change could be considered unreal. However, while they may be imaginary, they are not intangible, untrue or fake. They are real experiences for people, and they are not uncommon. Nearly one in five children surveyed by the BBC had had a bad dream about climate change.[32] While seemingly psychic, these experiences can be viscerally felt: Gergis grinds her teeth during her nightmares, for example. Further, these experiences are not a purely human cognitive achievement because they arise from our transcorporeal affective engagements with material worlds, whether that be with bushfire smoke or scientific documentation of the increasing frequency of such disasters. They also have significant relational effects on people, often alienating them from their friends and family, as some of the statements in this chapter's vignette indicate. Our climatically enmeshed bodies' mechanisms for holding tension and stress, and the entanglements of our cognitive knowledge, imaginary capacities, subconscious intelligences and somatic sensitivities mean that such experiences are both real and unreal; psychic and embodied; individually felt in specific moments and places, and globally and epochally dispersed.

While we could choose to dismiss these experiences, witnessing these un/realities is crucial for maintaining people's wellbeing and for cultivating more climate-responsible cultures. In the context of mass systemic denial and rapidly collapsing ecosystems, witnessing the realities of climate change can leave people feeling estranged from the business-as-usual world unfolding around them. For people to be able to fully engage with the material realities of climate change, we have to acknowledge that their surreal experiences are a legitimate part of the practice of witnessing global heating. To deny these bewildering affective experiences is to gaslight people, to dismiss their lived experience and deny their knowledges. That 'believing' in climate change can so frequently feel like living in an alternate universe demonstrates just how high the political, cultural, inter/personal and embodied stakes of doing so are. Listening to and legitimating people's un/real experiences of climate change is therefore a crucial political practice that enables people to affirm the validity of their concern about climate change.

Further, our affective and surreal experiences of climate change can enable us to experience and enact alternative ways of relating to and becoming (with) climate change. These ir/rational experiences can occur when we have loosened our cognitive attention and when our ability to think more relationally is potentially amplified (or less repressed).[33] This can enable us to feel 'multidimensional beings and space/times of the past-present-future...in one moment and in one place,'[34] such as when summer bushfires are glimpsed in the winter light. They enable us to combine the multiscalar but mechanistic knowledges of climate science with the connective and affective knowledges of our bodies' inter/personal, emplaced lived experiences, such as when the Pyrocene is experienced as the abandonment of burning children.[35] Rather than a system to be governed from afar by technologically empowered planetary managers, these un/realities enable us to experience climate change as something intimately entangled with our sense of self, our interpersonal relations, our moral responsibilities and our emplaced futures, which demands responding as part of that world.

Witnessing such climate un/realities could, if conducted carefully, increase settler people's capacity to witness Indigenous and animated climate cosmologies. The climate un/realities discussed earlier demonstrate that despite modernity's systemic objectification of the more-than-human world, climate can infiltrate, permeate, speak to and manifest in our supposedly individual human psyches. This is a glimpse, however preliminary, of climate's spiritual agency, where spirits arise from a 'participatory consciousness'[36] enacted through intimate, intergenerational, placed-based, multispecies relations. Examples of peoples whose cosmologies centre climate's spiritual agency include the Inuit, for whom *Sila* is a 'raw life force that lay[s] over the entire Land; that [can] be felt as air, seen as the sky, and lived as breath,'[37] who has 'intellectual, biological, psychological, environmental, locational, and geographical dimensions.'[38] For some Andean peoples, 'climate' includes powerful ancestral *ajayus,* spirits who embody the landscape, observe human people to check they are living morally sound lives, and control the weather accordingly.[39] This climate-human configuration is significantly distinct from the God's eye view of Western science where humans look upon and manage the climate. Dwelling longer with the notion of climate as a 'sentient commons,'[40] we might begin to appreciate that climate feels and is felt, knows and is known, speaks and is spoken to, in myriad ways.

To witness other climate realities does not require us to fully experience or know them. Rather, it is to accept them as real and valid despite our inabilities and differences. However, as Waanyi writer Alexis Wright points out, even settler people can and do appreciate, on some level, that weather and climate are ancient, inexplicable, uncontrollable and potentially spirited phenomena, so we are not beginning from complete incomprehension.[41] Witnessing Indigenous and animated climate realities is important because it is an issue of cognitive and ontological climate justice: 'whose knowledge is allowed to count as legitimate knowledge' and also 'whose reality is allowed to be real.'[42] This in turn matters because these realities are brought into being through intergenerational practices of multispecies interrelation; witnessing such realities enables these knowledge-making practices to continue to cultivate flourishing lifeworlds. However, because of the emphasis on rational truth, universally observable realities, and clearly measurable targets, mainstream climate advocacy often perpetuates the erasure of such animated climate knowledges.[43] Such objectification of reality is a feature of the extractivism that fuels climate change.[44] If settlers are to adequately address the climate crisis, we will need to move beyond the belief that we can comprehensively identify, measure, name and model every part of the climate through modernist rationality. Witnessing multiple climate realities can therefore enable an unsettling of extractivist realism, enroll us in non-anthropocentric relations and responsibilities, and open space for the realisation of more intimate, lived, climate worlds.

***

Witnessing climate change is fundamentally different to knowing about climate change. To witness climate change is to be called into affective relation, to be held accountable, and to be transformed by, climatic complexities. Through the various ways we can know *with, as* and *through* climate change, we enact distinct climate-human relationships, each of which contributes a partial perspective – a part, but not the entirety, of the vast and complex realities of climate collapse. There is no singular truth of a complete and universal climate reality. Even science denies as much as it explains, ignoring or overlooking the intricacies of our climatic entanglements.

Yet, climate change is demanding to be witnessed, and its affective weight surpasses science's best efforts to rationalise our knowledges of it. 'The' scientific reality can refract into seemingly unreal experiences. Everyday lived experiences demonstrate the complex and nuanced realities behind science's numbers, and animated climate cosmologies render apparent the anthropocentrism of science's God's eye view. Witnessing such multiple climate realities can help us better attune to the transcorporeal affective agency of climate change and articulate our relational climate responsibilities. Rather than seek conformity to one prescriptive 'correct' account, witnessing multiple climate realities can enable us to add additional modes of knowing, being and relating to climate change to our repertoires of climate responsivity.[45]

If we conceptualise knowing climate change as an affective labour of witnessing, we are better placed to adequately support people to begin and continue the traumatic-and-regenerative work of learning to live with climate change. We need to witness people as they witness climate change. Witnessing peoples' diverse climate realities validates them as legitimate climate knowers and ensures we do not gaslight them. This in turn can forge relationships built on trust and respect. In so doing, it can help us learn to live with each other in this changing climate and thus, it can catalyse alliances for collective climate action. In my climate change class, we were each witnessing climate change in diverse ways. In addition, through our class activities and discussions we were also witnessing each other as we encountered, witnessed and storied climate change. As I discuss in the next chapter, through these cascades of collective witnessing we crafted 'new stories about self, society and the world.'[46]

Notes

1 Dina Abbott and Gordon Wilson, *The Lived Experience of Climate Change: Knowledge, Science and Public Action* (Cham, Heidelberg, New York, Dordrecht, London: Springer International Publishing, 2015). Daniela Schofield and Femke Gubbels, "Informing Notions of Climate Change Adaptation: A Case Study of Everyday Gendered Realities of Climate Change Adaptation in an Informal Settlement in Dar Es Salaam," *Environment and Urbanization* 31, no. 1 (2019), https://doi.org/10.1177/095 6247819830074.

2 Debra Parkinson, "Investigating the Increase in Domestic Violence Post Disaster: An Australian Case Study," *Journal of Interpersonal Violence* 34, no. 11 (2019), https://doi.org/10.1177/0886260517696876.

3 Gemma Sou and Ruth Webber, "Disruption and Recovery of Intangible Resources During Environmental Crises: Longitudinal Research on 'Home' in Post-Disaster Puerto Rico," *Geoforum* 106 (2019).

4 Donna Haraway, "Situated Knowledges: The Science Question in Feminism and the Privilege of Partial Perspective," *Feminist Studies* 14, no. 3 (1988), https://doi.org/10.2307/3178066.

5 Deborah Bird Rose, *Reports from a Wild Country: Ethics for Decolonisation* (Sydney: University of New South Wales Press, 2004).

6 Louise Boscacci, "Wit(h)nessing," *Environmental Humanities* 10, no. 1 (2018).

7 James Felton, "This Australian Magpie Mimicking Emergency Sirens Is the Bleakest Thing You'll See Today," *IFL Science*, January 3, 2020. https://www.iflscience.com/plants-and-animals/this-australian-magpie-mimicking-emergency-sirens-is-the-bleakest-thing-youll-see-today/; Andy Moser, "Watch This Australian Magpie Perfectly Mimic the Sound of Emergency Sirens," *Mashable*, January 3, 2020. https://mashable.com/article/australian-magpie-mimic-siren-sound-wildfires/.

8 Haraway, "Situated Knowledges: The Science Question in Feminism and the Privilege of Partial Perspective."

9 Haraway, "Situated Knowledges: The Science Question in Feminism and the Privilege of Partial Perspective," 584.

10 Andrei Israel and Carolyn Sachs, "A Climate for Feminist Intervention: Feminist Science Studies and Climate Change," in *Research, Action and Policy: Addressing the Gendered Impacts of Climate Change*, ed. Margaret Alston and Kerri Whittenbury (Dordrecht, Heidelberg, New York and London: Springer, 2013).

11 Nancy Tuana, "Gendering Climate Knowledge for Justice: Catalyzing a New Research Agenda," in *Research, Action and Policy: Addressing the Gendered Impacts of Climate Change*, ed. Margaret Alston and Kerri Whittenbury (Dordrecht, Heidelberg, New York and London: Springer, 2013).

12 Mike Hulme, *Why We Disagree About Climate Change: Understanding Controversy, Inaction and Opportunity* (Cambridge and New York: Cambridge University Press, 2009), 3.

13 The rapidly emerging field of attribution climate science is increasing our capacity to conclude that specific weather events were (or were not) influenced by climate change, which is enabling those who are sticklers for scientific accuracy to acknowledge that experiences of weather can be experiences of climate change. Nonetheless, the exclusive acceptance of scientific modes of knowing and categorizing the world will limit our ability to accept and understand the *lived* experiences of such events, even when they are found, scientifically, to be experiences of climate change.

14 Lesley Head and Theresa Harada, "Keeping the Heart a Long Way from the Brain: The Emotional Labour of Climate Scientists," *Emotion, Space and Society* 24 (2017), https://doi.org/10.1016/j.emospa.2017.07.005.

15 Hannah Knox, "Thinking Like a Climate," *Distinktion: Journal of Social Theory* 16, no. 1 (2015): 98–9, https://doi.org/10.1080/1600910X.2015.1022565.

16 Lauren Rickards and James Watson, "Research Is Not Immune to Climate Change," *Nature Climate Change* 10, no. 3 (2020), https://doi.org/10.1038/s41558-020-0715-2.

17 For more discussion about the embodied nature of scientific practices of knowing climate, see Blanche Verlie, "Rethinking Climate Education: Climate as Entanglement," *Educational Studies* 53, no. 6 (2017), https://doi.org/10.1080/00131946.2017.1357555.

18 Haraway, "Situated Knowledges: The Science Question in Feminism and the Privilege of Partial Perspective," 586 & 83.

19 For a discussion on the persistent and problematic ableism of vision as a metaphor for knowledge, see McKnight and Whitburn below. Given the predominance of visual metaphors of climate knowledge, including in my students' comments, following Haraway and Hayward (below), I am using vision as a metaphor for climate knowledge here in recognition that there is no natural or normal way of seeing. Rather, all modes of vision are multi-sensorial and affective, and they are learned and enabled through more-than-human relations and hierarchies. Lucinda McKnight and Ben Whitburn, "The Fetish of the Lens: Persistent Sexist and Ableist Metaphor in Education Research," *International Journal of Qualitative Studies in Education* 30, no. 9 (2017), https://doi.org/10.1080/09518398.2017.1286407; Haraway, "Situated Knowledges: The Science Question in Feminism and the Privilege of Partial Perspective."; Eva Hayward, "Fingeryeyes: Impressions of Cup Corals," *Cultural Anthropology* 25, no. 4 (2010), https://doi.org/10.1111/j.1548-1360.2010.01070.x.

20 Nancy Tuana, "Viscous Porosity: Witnessing Katrina," in *Material Feminisms*, ed. Stacy Alaimo and Susan Hekman (Bloomington: Indiana University Press, 2008), 190.

21 Haraway, "Situated Knowledges: The Science Question in Feminism and the Privilege of Partial Perspective."

22 Israel and Sachs, "A Climate for Feminist Intervention: Feminist Science Studies and Climate Change."

23 The piece of writing used in our course was: Kari Marie Norgaard, "Climate Denial: Emotion, Psychology, Culture, and Political Economy," in *The Oxford Handbook of Climate Change and Society*, ed. John Dryzek, Richard Norgaard, and David Schlosberg (New York: Oxford University Press, 2011).

24 Paul Gilding, "The Parallel Universes of Climate Change. Where Do You Live?," *Paul Gilding*, 2009, https://paulgilding.com/2009/09/10/cc2009091 0paralleluniverses/.

25 For example: Dana Nuccitelli, "Climate Denial Is Like the Matrix; More Republicans Are Choosing the Red Pill," *The Guardian*, July 19, 2017, https://www.theguardian.com/environment/climate-consensus-97-percent/2017/jul/19/climate-denial-is-like-the-matrix-more-republicans-are-choosing-the-red-pill

26 The Wachowskis, "*The Matrix*," (United States of America: Warner Bros., 1999), Motion picture. In *The Matrix* machines have taken control of the planet and they breed human bodies to create electrical energy to power themselves. The Matrix is a computer-generated simulation designed and operated by the machines which humans are plugged into so that the machines can pacify and thus farm them. In *The Matrix* (the film), after

becoming aware that reality is not what it seems, the protagonist Neo is offered a choice between a red pill or a blue pill: if he chooses the blue pill, he will go back to obliviously living in the Matrix (the comfortable computer simulation); if he chooses the red pill, he will find out the truth and live in the real but distressing world. Notably, only by living in reality will he be able to contribute to a better real world.

27 David Johnson, "The Meditation of the Red Pill," *Transition Voice* (2011). http://transitionvoice.com/2011/02/the-meditation-of-the-red-pill/.

28 Karen Barad, *Meeting the Universe Halfway: Quantum Physics and the Entanglement of Matter and Meaning* (Durham and London: Duke University Press, 2007); Karen Barad, "Diffracting Diffraction: Cutting Together-Apart," *Parallax* 20, no. 3 (2014), https://doi.org/10.1080/13534645.2014.927623; Donna Haraway, *Modest−Witness@Second−Millennium.Femaleman−Meets−Oncomouse: Feminism and Technoscience* (New York and London: Routledge, 1997).

29 Brigid Delaney, "2019 Wasn't Just Protests and Fleabag: It Was the Year a Climate Truth Bomb Dropped," *The Guardian*, December 20, 2019. https://www.theguardian.com/commentisfree/2019/dec/19/2019-wasnt-just-protests-and-fleabag-it-was-the-year-a-climate-truth-bomb-dropped. (Original emphasis).

30 Stefan Rahmstorf, "This Is How Scientists Feel," *Is This How You Feel?* (2017). https://www.isthishowyoufeel.com/this-is-how-scientists-feel.html.

31 Joëlle Gergis, "Witnessing the Unthinkable," *The Monthly*, July 2020. https://www.themonthly.com.au/issue/2020/july/1593525600/jo-lle-gergis/witnessing-unthinkable#mtr.

32 Richard Atherton, "Climate Anxiety: Survey for BBC Newsround Shows Children Losing Sleep over Climate Change and the Environment," *BBC*, March 3, 2020. https://www.bbc.co.uk/newsround/51451737.

33 Kirsten Anker, "Law as...Forest: Eco-Logic, Stories and Spirits in Indigenous Jurisprudence," *Law Text Culture* 21, no. 1 (2017).

34 Felicity Amaya Schaeffer, "Spirit Matters: Gloria Anzaldúa's Cosmic Becoming across Human/Nonhuman Borderlands," *Signs: Journal of Women in Culture and Society* 43, no. 4 (2018): 1006, https://doi.org/10.1086/696630.

35 For more discussion of the ways scientific and embodied experience can combine to cultivate such multitemporal and global-personal phenomena, see Astrida Neimanis, *Bodies of Water: Posthuman Feminist Phenomenology* (London: Bloomsbury Academic, 2017).

36 Anker, "Law as...Forest: Eco-Logic, Stories and Spirits in Indigenous Jurisprudence," 194.

37 Rachel Qitsualik cited in Zoe Todd, "An Indigenous Feminist's Take on the Ontological Turn: 'Ontology' Is Just Another Word for Colonialism," *Journal of Historical Sociology* 29, no. 1 (2016): 5, https://doi.org/10.1111/johs.12124.

38 Rachel Qitsualik, "Inummarik: Self-Sovereignty in Classic Inuit Thought," in *Nilliajut: Inuit Perspectives on Security, Patriotism and Sovereignty* (Ottawa: Inuit Qaujisarvingat, 2013), 29.

39 Anders Burman, "The Political Ontology of Climate Change: Moral Meteorology, Climate Justice, and the Coloniality of Reality in the

Bolivian Andes," *Journal of Political Ecology* 24, no. 1 (2017). https:// doi.org/10.2458/v24i1.20974.
40 Todd, "An Indigenous Feminist's Take on the Ontological Turn: 'Ontology' Is Just Another Word for Colonialism," 20.
41 Alexis Wright, "Deep Weather," *Meanjin* 2 (2011). https://meanjin.com.au/ essays/deep-weather/.
42 Burman, "The Political Ontology of Climate Change: Moral Meteorology, Climate Justice, and the Coloniality of Reality in the Bolivian Andes," 925.
43 Bawaka Country et al., "Gathering of the Clouds: Attending to Indigenous Understandings of Time and Climate through Songspirals," *Geoforum* 108 (2020), https://doi.org/10.1016/j.geoforum.2019.05.017; Burman, "The Political Ontology of Climate Change: Moral Meteorology, Climate Justice, and the Coloniality of Reality in the Bolivian Andes."
44 Hugo Reinert, "About a Stone: Some Notes on Geologic Conviviality," *Environmental Humanities* 8, no. 1 (2016), https://doi.org/10.1215/2201191 9-3527740.
45 Bruno Latour, "How to Talk About the Body? The Normative Dimension of Science Studies," *Body and Society* 10, no. 2–3 (2004), https://doi.org/1 0.1177/1357034X04042943.
46 Sally Gillespie, "Climate Change and Psyche: Conversations with and through Dreams," *International Journal of Multiple Research Approaches* 7, no. 3 (2013): 344, https://doi.org/10.5172/mra.2013.7.3.343.

References

Abbott, Dina, and Gordon Wilson. *The Lived Experience of Climate Change: Knowledge, Science and Public Action.* Cham, Heidelberg, New York, Dordrecht, London: Springer International Publishing, 2015.
Anker, Kirsten. "Law as...Forest: Eco-Logic, Stories and Spirits in Indigenous Jurisprudence." *Law Text Culture* 21, no. 1 (2017): 191–213.
Atherton, Richard. "Climate Anxiety: Survey for Bbc Newsround Shows Children Losing Sleep over Climate Change and the Environment." *BBC* (March 3 2020). https://www.bbc.co.uk/newsround/51451737.
Barad, Karen. "Diffracting Diffraction: Cutting Together-Apart." *Parallax* 20, no. 3 (2014): 168–87. https://doi.org/10.1080/13534645.2014.927623.
Barad, Karen. *Meeting the Universe Halfway: Quantum Physics and the Entanglement of Matter and Meaning.* Durham and London: Duke University Press, 2007.
Bawaka Country, S. Wright, S. Suchet-Pearson, K. Lloyd, L. Burarrwanga, R. Ganambarr, M. Ganambarr-Stubbs, B. Ganambarr, and D. Maymuru. "Gathering of the Clouds: Attending to Indigenous Understandings of Time and Climate through Songspirals." *Geoforum* 108 (2020): 295–304. https:// doi.org/10.1016/j.geoforum.2019.05.017.
Boscacci, Louise. "Wit(h)nessing." *Environmental Humanities* 10, no. 1 (2018): 343–47.
Burman, Anders. "The Political Ontology of Climate Change: Moral Meteorology, Climate Justice, and the Coloniality of Reality in the Bolivian

Andes." *Journal of Political Ecology* 24, no. 1 (2017): 921–30. https://doi.org/10.2458/v24i1.20974.

Delaney, Brigid. "2019 Wasn't Just Protests and Fleabag: It Was the Year a Climate Truth Bomb Dropped." *The Guardian*, December 20, 2019. https://www.theguardian.com/commentisfree/2019/dec/19/2019-wasnt-just-protests-and-fleabag-it-was-the-year-a-climate-truth-bomb-dropped.

Felton, James. "This Australian Magpie Mimicking Emergency Sirens Is the Bleakest Thing You'll See Today." *IFL Science*, January 3, 2020. https://www.iflscience.com/plants-and-animals/this-australian-magpie-mimicking-emergency-sirens-is-the-bleakest-thing-youll-see-today/.

Gergis, Joëlle. "Witnessing the Unthinkable." *The Monthly*, July 2020. https://www.themonthly.com.au/issue/2020/july/1593525600/jo-lle-gergis/witnessing-unthinkable#mtr.

Gilding, Paul. "The Parallel Universes of Climate Change. Where Do You Live?" *Paul Gilding*, 2009. https://paulgilding.com/2009/09/10/cc20090910paralleluniverses/.

Gillespie, Sally. "Climate Change and Psyche: Conversations with and through Dreams." *International Journal of Multiple Research Approaches* 7, no. 3 (2013): 343–54. https://doi.org/10.5172/mra.2013.7.3.343.

Haraway, Donna. *Modest−Witness@Second−Millennium.Femaleman−Meets−Oncomouse: Feminism and Technoscience*. New York and London: Routledge, 1997.

Haraway, Donna. "Situated Knowledges: The Science Question in Feminism and the Privilege of Partial Perspective." *Feminist Studies* 14, no. 3 (1988): 575–99. https://doi.org/10.2307/3178066.

Hayward, Eva. "Fingeryeyes: Impressions of Cup Corals." *Cultural Anthropology* 25, no. 4 (2010): 577–99. https://doi.org/10.1111/j.1548-1360.2010.01070.x.

Head, Lesley, and Theresa Harada. "Keeping the Heart a Long Way from the Brain: The Emotional Labour of Climate Scientists." *Emotion, Space and Society* 24 (2017): 34–41. https://doi.org/10.1016/j.emospa.2017.07.005.

Hulme, Mike. *Why We Disagree About Climate Change: Understanding Controversy, Inaction and Opportunity*. Cambridge and New York: Cambridge University Press, 2009.

Israel, Andrei, and Carolyn Sachs. "A Climate for Feminist Intervention: Feminist Science Studies and Climate Change." In *Research, Action and Policy: Addressing the Gendered Impacts of Climate Change*, edited by Margaret Alston and Kerri Whittenbury, 33–51. Dordrecht, Heidelberg, New York and London: Springer, 2013.

Johnson, David. "The Meditation of the Red Pill." *Transition Voice*, 2011. http://transitionvoice.com/2011/02/the-meditation-of-the-red-pill/.

Knox, Hannah. "Thinking Like a Climate." *Distinktion: Journal of Social Theory* 16, no. 1 (2015): 91–109. https://doi.org/10.1080/1600910X.2015.1022565.

Latour, Bruno. "How to Talk About the Body? The Normative Dimension of Science Studies." *Body and Society* 10, no. 2–3 (2004): 205–29. https://doi.org/10.1177/1357034X04042943.

McKnight, Lucinda, and Ben Whitburn. "The Fetish of the Lens: Persistent Sexist and Ableist Metaphor in Education Research." *International Journal of Qualitative Studies in Education* 30, no. 9 (2017): 821–31. https://doi.org/10.1080/09518398.2017.1286407.

Moser, Andy. "Watch This Australian Magpie Perfectly Mimic the Sound of Emergency Sirens." *Mashable*, January 3, 2020. https://mashable.com/article/australian-magpie-mimic-siren-sound-wildfires/.

Neimanis, Astrida. *Bodies of Water: Posthuman Feminist Phenomenology.* London: Bloomsbury Academic, 2017.

Norgaard, Kari Marie. "Climate Denial: Emotion, Psychology, Culture, and Political Economy." In *The Oxford Handbook of Climate Change and Society*, edited by John Dryzek, Richard Norgaard and David Schlosberg, 399–413. New York: Oxford University Press, 2011.

Nuccitelli, Dana. "Climate Denial Is Like the Matrix; More Republicans Are Choosing the Red Pill." *The Guardian*, July 19, 2017. https://www.theguardian.com/environment/climate-consensus-97-per-cent/2017/jul/19/climate-denial-is-like-the-matrix-more-republicans-are-choosing-the-red-pill.

Parkinson, Debra. "Investigating the Increase in Domestic Violence Post Disaster: An Australian Case Study." *Journal of Interpersonal Violence* 34, no. 11 (2019): 2333–62. https://doi.org/https://doi.org/10.1177/0886260517696876.

Qitsualik, Rachel. "Inummarik: Self-Sovereignty in Classic Inuit Thought." In *Nilliajut: Inuit Perspectives on Security, Patriotism and Sovereignty*, edited by Scot Nickels, Karen Kelley, Carrie Grable, Martin Lougheed, and James Kuptana, 23–34. Ottawa: Inuit Tapiriit Kanatami, 2013.

Rahmstorf, Stefan. "This Is How Scientists Feel." *Is This How You Feel?* (2017). https://www.isthishowyoufeel.com/this-is-how-scientists-feel.html.

Reinert, Hugo. "About a Stone: Some Notes on Geologic Conviviality." *Environmental Humanities* 8, no. 1 (2016): 95–117. https://doi.org/10.1215/22011919-3527740.

Rickards, Lauren, and James Watson. "Research Is Not Immune to Climate Change." *Nature Climate Change* 10, no. 3 (2020): 180–83. https://doi.org/10.1038/s41558-020-0715-2.

Rose, Deborah Bird. *Reports from a Wild Country: Ethics for Decolonisation.* Sydney: University of New South Wales Press, 2004.

Schaeffer, Felicity Amaya. "Spirit Matters: Gloria Anzaldúa's Cosmic Becoming across Human/Nonhuman Borderlands." *Signs: Journal of Women in Culture and Society* 43, no. 4 (2018): 1005–29. https://doi.org/10.1086/696630.

Schofield, Daniela, and Femke Gubbels. "Informing Notions of Climate Change Adaptation: A Case Study of Everyday Gendered Realities of Climate Change Adaptation in an Informal Settlement in Dar Es Salaam."

Environment and Urbanization 31, no. 1 (2019): 93–114. https://doi.org/10.11 77/0956247819830074.

Sou, Gemma, and Ruth Webber. "Disruption and Recovery of Intangible Resources During Environmental Crises: Longitudinal Research on 'Home' in Post-Disaster Puerto Rico." *Geoforum* 106 (2019): 182–92.

The Wachowskis. *The Matrix.* United States of America: Warner Bros., 1999. Motion picture.

Todd, Zoe. "An Indigenous Feminist's Take on the Ontological Turn: 'Ontology' Is Just Another Word for Colonialism." *Journal of Historical Sociology* 29, no. 1 (2016): 4–22. https://doi.org/10.1111/johs.12124.

Tuana, Nancy. "Gendering Climate Knowledge for Justice: Catalyzing a New Research Agenda." In *Research, Action and Policy: Addressing the Gendered Impacts of Climate Change*, edited by Margaret Alston and Kerri Whittenbury, 17–31. Dordrecht, Heidelberg, New York and London: Springer, 2013.

Tuana, Nancy. "Viscous Porosity: Witnessing Katrina." In *Material Feminisms*, edited by Stacy Alaimo and Susan Hekman, 188–213. Bloomington: Indiana University Press, 2008.

Verlie, Blanche. "Rethinking Climate Education: Climate as Entanglement." *Educational Studies* 53, no. 6 (2017): 560–72. https://doi.org/10.1080/00131 946.2017.1357555.

Wright, Alexis. "Deep Weather." *Meanjin* 2 (2011): 70–82. https://meanjin.com.au/essays/deep-weather/.

5   Storying climate collectives

Climatechange is unrelenting and real, a crazy ethical labyrinth which may lead to a devastating conclusion.

The climate is changing because of human designed and constructed processes which perpetuate the unsustainable extraction, production and disposal of goods that we have based our whole lives around.

It is a poor choice to wait for a new destructive level of nature to manifest itself before we (globally) start to take action on climate change. But this may be the only way to inspire such actions, as humans are woeful at responding to slow moving dangers. This is not to say I am entirely cynical about humanity's chances of surviving this growing threat. There are countless institutions and communities actively reducing their impacts on the world and its natural systems. My totally cynical view is that non-fossil-fuel-based energy production will only become the norm once the renewable-energy corporations can provide more money than fossil fuel corporations in bribes to political interests.

On the other hand, it's possible that, in the deep future, the effects of humanity on the climate today will be negligible, perhaps immeasurable when compared against the larger state of the Earth where the climate changes dramatically over the course of millions of years. To believe that we, today, can have such large scale, long term impacts on the Earth admittedly seems quite arrogant: the Earth and its life systems have mitigated and adapted to threats larger than anthropogenic climate change for billions of years.

But we still need to ask, how can we as humans confront the ethical dilemmas that planetary management demands of us?

DOI: 10.4324/9780367441265-5

Ultimately, I think that humans are not inherently lazy, or inherently as apathetic as we are today. I think we are products of our direct environments, and we have tailored our environments to such a degree that comfort numbs our will to act on our beliefs. The ruling decisions of the human race are centred around short-term profit, focussing on self-interest and money over the natural environment. This capitalist structure also renders those towards the lower-middle income bracket far more at risk to the impacts of climate, while leaving them with far less resilience to cope.

These political and economic structures that underpin society have led to my cynicism.

I feel this frustration towards governments, companies, humans and myself.

This is what intrigued me the most, the aspect of neglect and the level of response, or lack thereof, in the face of the most significant environmental issue of our time. I wonder what truly is the implication of a species that conducts themselves in the manner in which we do; is our current environmental situation a product of our arrogance?

Our ignorance?

Our indifference?

Our apathy?

Our greed?

Our hubris, or all of the above?

Climate change not only destroys some of the most beautiful and vulnerable places in the world but also ancient living cultures. It is impacting Indigenous land and cultures, cultures which in my opinion may be one of the keys in combating the problem, through a deeper respect for the land and for each other.

There are limited grand visions of what a positive low- or zero-carbon world may actually look like, no decisive government plans or political leadership towards a vision of a reinvented future. The discourse is often around giving something up or losing the cultural benefits which have created who we are, as opposed to transforming them. The pursuit of liberal freedom is often misrepresented by the false pretence that market freedom equates to individual freedom and choice. But what choices are we not being offered through this model and who is in fact losing their freedom? What about the benefits of a new vision and global change?

If we as young Australians look to Indigenous sustainable land use practices, by changing the way we understand climate, we could change our interactions with climate. We could shape our own national identity around climate, and create a culture built on shared values and vision to accelerate global behavioural change.

Listening to students in class and all the guest speakers was really interesting as they all came from such different areas, yet they are all working on the one thing: climate change. I've felt inspired by their contributions. They have progressed my understanding of climate change, sure, but they've also offered a view to what is at times the front lines of climate change responses. Despite the sometimes gloomy subject matter, I've often left lectures feeling hopeful, for I can begin to envisage the path ahead, and also how I might have a role in creating it.

One day after class, I felt like I was floating on the way home. Maybe I was delirious because this subject matter is so exhausting. But I really felt buoyed by the energy everyone brings to class.

Through participating in this class, I feel inspired to move forward, to further develop my understanding of climate change and share my newly acquired knowledge.

But the ways that governments, large corporations and individuals continue to act, in the face of already visible consequences of human impact, still causes me great stress and confusion. I am left with an overwhelming sense of guilt, and many questions surrounding the basic morality of humankind which sees so many people turn a blind eye.

***

Climate change communication is increasingly recognising the powerful role that stories play in cultivating engagement with climate change. For example, research finds that stories of the impacts of climate change can help make it feel relevant and important, and stories of people taking action can help doing so seem achievable and effective.[1] Whether true or fictional, stories of climate change can enable us to empathise with others' experiences and help us become more considerate of different perspectives. They can transport us to seemingly distant times and places, enabling us to situate our lives and actions in complex histories and planetary futures.

Stories connect occurrences and information into meaningful and memorable narratives. Stories are powerful social technologies because they are connective and affective: they articulate relations between beings, places and things, and they activate our emotions far more easily than statistics and graphs. Stories are also engaging because they often address issues or topics from relatable perspectives such as individual experiences. In order to be engaging, stories have to be socially comprehensible and they thus embed and enact beliefs about ontology (what exists), subjectivity and embodiment (the 'who' of the story), causality (why and how things happen), agency (who or what can effect change) and temporality. To be socially comprehensible, these beliefs have to align with or build upon established ones, but they can also rework them to some extent. Stories thus rehearse, but can also regenerate, social scripts and templates.[2] This makes telling stories a dynamic social practice which works in multiple ways to influence what we think is possible, likely and desirable, and which emerges from and actively reconfigures more-than-human worlds.[3]

For such reasons, climate change communicators, whether documentary makers, fiction writers, journalists, educators or activists, often use personal stories – sometimes their own, sometimes those of others – to engage their audiences. We tell stories from the perspectives of individual humans because this helps people translate global heating into realms of reality that resonate with their lived experience. For example, the loss of three billion animals in Australia's 2019/2020 bushfires[4] is shocking – statistics are not always tedious – but it is made all the more real when partnered with the first-hand accounts of beekeepers traumatised by animals screaming in the bush and footage of wildlife carers treating singed marsupials.[5] Documenting how specific humans experienced and responded to the realities behind that statistic provides a relatable perspective that enables us to process what it means, why it matters, and what we could do about it. As evidence of this, wildlife carers were inundated with new volunteer applications following the 2019/2020 bushfires.[6] However, might it be possible that stories of individual people in discrete events may reinforce our inability to think and relate to climate change's slower and non-human emergencies, like sea level rise and mass extinction? Could putting a human face on climate change perversely re-centre anthropocentric individuality?

Debates in climate change fiction circles complicate the idea that stories of individual experiences of climate change will address all the challenges of engaging people in this pressing and overwhelming issue. Those most skilled at the arts of composing engaging stories and interrogating their social agency – fiction writers and literary critics – are questioning whether individuated human perspectives are up to the task of narrating the monumental planetary warping that climate change engenders.[7] Climate change operates over multigenerational timescales and both arises from and demands collective human action; so, 'to understand climate change one needs to go beyond normal[ised] human experience.'[8] Yet individualised personal stories are both comprehensible and highly engaging, leaving us facing 'a crisis of narrative.'[9] Yes, communicators need to make the globalised science relatable. But we also need people to become better able to relate to patterns, processes and cycles that operate beyond the realm of immediate individual human experience.[10] To do this, we need stories that attune to more collective, transcorporeal, geographically distributed and multi-temporal experiences and occurrences.[11] If we, collectively, are best engaged by stories that are relatable to our sense of self, then we need stories that can expand this such that long term, non-linear and distant events and relations become personally salient. We need stories that transform our understanding of what it is to be human, stories that dissolve, or at least blur, the skin-bound individual as the template for relatability.

The genre of climate fiction is rapidly diversifying and experimenting with such considerations, playing with temporalities, plots and characters in order to better represent the 'derangement' of climate change in relatable ways.[12] However, no matter how much climate fiction might reconfigure its protagonists towards more collective, distributed, and more-than-human forms, these stories are conceptualised to be told by individuated authors to individuated readers. Both climate change communication and climate fiction typically enact a monologue with their audience, whereby clever experts carefully develop a story to be distributed to relatively passive audiences. But part of the problem that we face in climate change engagement is not that most people do not hear about climate change, it is that most people do not talk about climate change. Most of us are consumers, not creators, of climate stories, with scientists, politicians, deniers and the media offering competing monologues for the docile consumption of the public.[13]

There are numerous problems in this one-way human-centred model of story production and dissemination. The idea that certain people have the capacity to 'frame' climate change as though it can be neatly contained in an ideological box of the communicator's choice perpetuates the myth that if we humans just put enough effort in, we can competently oversee how the world unfolds – in this case through exerting discursive power over other humans, who, once they are effectively engaged on the issue, will in turn act on the climate. Further, while people are of course engaged by stories told by others, we know that people are most engaged by stories that they are actively involved in creating themselves.

If we are to learn to live with climate change, we need people to be personally engaged in the collective composition of stories that radically reimagine human-climate relations. We also need to ensure that how we conceptualise the practice of storying climate change does not reinstate anthropocentric notions of authorship as a form of authority and control. This chapter thus attunes to the ways that climate change's affective agency participates in the kinds of stories we can and do tell about collectives of humans, and thus, how climate change contributes to the ways we are becoming human. Storying-with my climate-changed class, I suggest that our narrative practices contributed to us becoming a 'cloudy collective': a moody, ephemeral, more-than-human ensemble that participated in and emerged from our changing climate. This temperamental, shape-shifting, expansive 'self' that emerged in our class is also storied into this book as the narrator of the vignettes. You might re-read them and notice, amidst the generally cohesive and coherent narrative: moments of disjuncture; competing and contrasting philosophies and experiences; oscillation between 'I' and 'we' statements as students narrated their own identification with and alienation from others; and how frequently these practices of attachment and disconnection are enacted through affective similarities and differences. This is a story of affective transformation in progress: the overwhelming distress of climate change was unravelling who we were, and our practices of expressing these emotions were contributing to our ongoing recomposition.

***

Stories of collectives of humans are some of the most powerful climate stories we tell, because our shared climatic futures are co-composed by things collectives of humans do and do not do. Beliefs about human traits and capacities influence what we think is possible and thus, worth investing our energy in. At the same time, we know that people's emotions play crucial roles in the ways they do, or do not, engage with climate change, with both concern and hope (among others) having important roles to play in motivating and sustaining engagement. Rarely are climate change's collective nature and its intense emotionality considered together, however, doing so can help craft more hopeful climate stories. The inevitable answer to 'what can I do about climate change?' is 'not much,' or at least, 'not enough,' a conclusion that can spur pessimism and even excuse us from responsibility. This is especially prevalent in cultures seeped in narratives of individualistic entrepreneurs and superheroes who solve their own and everyone else's problems all by themselves; if we cannot successfully achieve something by ourselves, we reason that we cannot do anything at all.

When I hear people say they feel hopeful (or otherwise energised) regarding climate change, they almost always express these feelings in relation to the actual or potential establishment of climate-active groups of humans. Knowing that others share their frustrations, care about the world, and are doing things to address climate change provides inspiration, motivation and conviction that we can turn things around. However, this can be problematic if the collective in which hope is located is considered to be external to the self, because such narratives become an excuse to sit by and wait for someone else to step up. As Greta Thunberg, articulating many young people's sentiments, stated at the European Economic and Social Committee, in 2019:

> People always tell us that they are so hopeful. They are hopeful that the young people are going to save the world, but we are not. There is simply not enough time to wait for us to grow up and become the ones in charge...You can't just sit around waiting for hope to come...Hope is something you have to earn.[14]

For collectives to really generate hope, they must be collectives that people can join, or that enable them to participate in some way. Distinct from those Thunberg refers to who are soothed, but not activated, by youth climate action, the following student's end of semester reflection heralds an all-encompassing collective that includes themself as part of the 'we.' This story is more promising because it therefore enrols them in the hope-generating collective, and thus, in the action required:

Through all this overwhelming, devastating news and under-standing the reality of climate change, I have grasped hold of an exciting potential for global unification. I have been encouraged by the possibility that the climate changing might be the one factor that pulls our divided world together to form a united social mass of individuals who want to see a brighter future and to see people and our environment being put over profit...I hope that climate change becomes a reality we work with and respect.

Stories of inclusive collectives are more encouraging for these reasons: inscribing people into an active collective affects how they understand and conduct themselves, and it can therefore bring collective action into reality through rehearsing it.

However, the most seemingly inclusive collective, notions of a homogeneous 'humanity,' tend to suffocate hope rather than inspire it; the student's comment above is a speculative example, dependent on a unified humanity that does not yet exist. Consider some of the statements in this chapter's vignette: when it comes to climate change, the collective character of 'humanity' is often composed through presuming that the features of neoliberal-colonial systems, such as self-interest, are universal features of human DNA.[15] When contemporary industrial practices are inferred to be essential traits of 'the human race,' our outlook is grim, as one of my student's statements indicates: 'human nature can be cruel, unforgiving and damaging to any form of sustained life.' When we associate such qualities as essential characteristics of humans, we lead ourselves to conclusions such as this student did: 'as long as humans dominate the world then there is no hope.'

This inaccurate and defeatist logic is embedded in one of the on-trend ways of discussing climate change and other widescale ecological crises. The *Anthropocene* is a term proposed to denote the emerging geological epoch – 'the age of man' – that will be characterised by mass ecosystem collapse, mass extinction, rapid planetary heating, and, depending on who you speak to, civilisational collapse and even the extinction of humans (ironically enough). The premise is that the material legacies of these ruptures will be deposited in the Earth's crust as geological layers of compressed toxins and extinct species. This notion is alluring because it seems to elicit concern about the ecological damage 'humans' are causing. Yet its appeal is also due in part to its narcissism which trades in self-aggrandizing notions of human agency.[16] This term's comprehensibility is achieved by normalising and promoting a very specific version of the human, typically an entrepreneurial, white, able-bodied, heterosexual, male individual whose economic rationalism inevitably leads to planetary destruction.[17] This erases, and also dehumanises, anyone who does not fit into this category: Indigenous people, poor people, those engaged in caring labour and the billions of people with negligible carbon footprints, among others.[18] Because of this, the notion of the Anthropocene fails to provide any vision for preventing its own emergence. In this way it is a self-fulfilling prophecy that both decries and summons a world in which humans catalyse the rapid collapse of Earth's life systems. The Anthropocene is a term that strives to motivate change but falls flat because of its inaccurate and uninspiring assumptions about 'humanity.'[19]

Individualising climate responsibility and misanthropic notions of human collectives both lead us to self-perpetuating despair. This is because they are in fact the same story, just operating at different scales. The Anthropocene is a story narrating the aggregate agency of anthropocentric individualism, where masses of cookie-cutter individual humans pollute the atmosphere and fail to cooperate to address this. Storying climate change as 'human induced' is discouraging because it operates through normalising problematic ways of being human and assimilating human diversity into this toxic character.[20] This false unity is severely oppressive and depressing, excluding, and even erasing, multitudes while perpetuating anthropocentrism's victimisation of the non-human world. Rather than look to existing, and potential, alternative ways of be(com)ing human, the Anthropocene story is literally sedimenting anthropocentrism into the Earth's strata, which is what makes it so petrifying.

***

If stories of individuals are disheartening, and stories that homogenise humans tend to normalise extractivism, what sorts of stories might encode and envision the diverse ways of being, relating and responding that are necessary for climate justice? In our class, students referred to a group that emerged through the affective atmosphere composed by witnessing each other's encounters with climate change. As discussed in Chapters 3 and 4, engaging in the labour of witnessing climate change had generated climate anxiety, which countered our individualistic subjectivities. As we collectively narrated our experiences of encountering and witnessing climate change's affectivity, an indistinct yet undeniable collective emerged that enabled us to 'stay with the trouble'[21] of climate vulnerability and complicity. Subsequently, students began speaking about this intangible group, and the stories they told generated different feelings and identifications which enabled them to break down the myth of *Homo destructivus*. Drawing upon my own affective entanglement in this congregation, I suggest it can be storied as a 'cloudy collective,' one example of the kinds of diverse, transformational, non-anthropocentric characters we need for cultivating the capacities to learn to live with climate change.

Clouds are an apt figure for storying more-than-human climate capable collectives. Clouds are gatherings of liquid or solid matter suspended in an atmosphere, and 'cloud' can also refer to other kinds of collectives: a cloud of electrons, or a cloud of gnats, for example. Clouds are aggregations that arise from the relations between the ecological, hydrological, atmospheric, geological and social, and form depending on processes spanning the molecular to the planetary, the momentary to the epochal. The multiple forces which compose clouds are themselves constantly in flux, and thus while clouds have some sense of togetherness they are constantly metamorphosing. As such, clouds do not have clear boundaries or edges, but rather, a kind of wispy transition zone where the assemblage coalesces, extends, transforms, retracts, and/or dissipates through its relationship with the wider atmosphere. This permeability means clouds are constantly incorporating and excluding particular parts of the world according to particular energetic flows. Cloudiness also obscures visibility, and it thus attends to the difficulty of identifying and delineating exactly how, where, when and by whom such groups are composed. Cloudy collectives are therefore shifty: changeable and hard to pin down. They are averse to unity, stability and definability, yet this enables them to be more inclusive and expansive, even if we cannot fully trace these movements.

Clouds are more than a metaphor for human collectives. We are not just 'like' clouds. As breathing, sweating, radiating bags of gas and liquid that metabolise and reconfigure carbon, hydrogen and oxygen, human bodies are 'only precariously contained in a skin sac.'[22] We are 'instead profoundly distributed, inherited, gestational [and] differentiated' bodies of water,[23] and we are implicated in the existence of clouds as much as they precipitate the conditions that make our lives possible. We are grounded and encased clouds, meaty interfaces that contribute to our floating counterparts' emergence in multiple ways. The humid air we expire is but a breath of cloud, and one that percolates through our multispecies relations in its journey of planetary circulation and weathering. As climate-changers, the emissions we produce form clouds (of a kind), and these emissions' planetary lives in turn contribute to the cloudiness, or lack thereof, of the atmosphere. Rainclouds soothe drought and fire-stricken country, but also produce floods, hail and blizzards; and clouds can both reflect the sun's radiation or trap its energy here on earth. Clouds, therefore, are gaseous extensions of our earth-bound bodies, and just like those of us who are climate-concerned-and-complicit, their roles in planetary heating are ambiguous.

Related to their atmospheric transcorporeality, cloudy collectives are profoundly affective: they are composed through circulating earthly energies, including climate anxiety. As The Cloud Appreciation Society put it, 'clouds are expressions of the atmosphere's moods.'[24] Accordingly, in a climate-changing world, cloudy collectives are temperamental, emerging from and contributing to gloom, frustration and heaviness, yet also lightness, pleasure and refreshment. These moody menageries emerge through, and in turn stimulate, our breathy practices of collectively storying climate change. Cloudy collectives are composed as our voices crack when we verbalise the violences of climate injustice; as we groan with exasperation at governments approving new fossil fuel projects; as we whisper our fears in climate grief workshops; as we shout 'climate action now' at rally after rally, after rally; and as we wheeze on the phone trying to convey the horror of bushfire smoke to distant relatives.

Cloudy collectives are distinctly non-anthropocentric and more-than-human. Climate change is not just something they organise themselves around in order to act on it. Rather, climate change is a key participant in the composition of cloudy collectives. Our complicity in and vulnerability to climate change generates conflicting feelings of outrage, despair, guilt and tentative hope, and those feelings can both connect and alienate us from others. Climate change's affective agency influences how we feel and who we do and do not identify with, which makes it a central player in the coalitions we form. To be a little more precise, climate change, our feelings and our relations co-compose each other. Attuning to the ways these climatic-affective intensities flow through and compose our transcorporeal collectives debunks both the autonomous individual and the homogenised humanity of the Anthropocene story. It curates an appreciation of the ways that climate change acts on us, including through the stories we are enabled and compelled to tell of it.

The Bawaka Country research collective offers one detailed account of a kind of cloudy collective composed through storying-with climate. *Wukun* or 'gathering of the clouds' is a Yolŋu songspiral. Yolŋu are Indigenous Australians from northeast Arnhem land. Songspirals are intergenerational, deeply emplaced, relational practices that 'make and remake the life-giving connections between people and place' and thus 'sing Country and time into existence through co-becoming.'[25] Wukun̲ is enacted by walking along and with Country, and calling to the clouds, which gather in response and in so doing, gather the people. Wukun̲ articulates and enacts a more-than-human collectivity that is responsive to climate, and whose responsibility is enacted through storying its intimate relationality with climate. This collective is also multi-temporal and intergenerational:

> for Yolŋu, the singer does not simply exist as a lone individual, acting in a single time. The singer is...those that sing together...the clan whose homeland it is...people who have passed away both recently and a long time ago...those of the clan who will pass away in the future [and] the clouds that gather.[26]

Wukun̲, the songspiral which stories generations of Yolŋu into relation with Country, involves an 'attention that is richly affective and comes from deep, embodied knowledge – seeing, hearing, feeling, laughing.'[27] Through walking-, singing- and dancing-with clouds, Yolŋu become-with the clouds and 'join to form an enormous cloud ready for the rain.'[28] In these ways, these more-than-verbal practices of lived storytelling enact a world in which clouds, skies, weather, climate, place and people are responsive to, for, with and as each other. In contrast to the Anthropocene narrative, Wukun̲ offers us a story of climate collectives where people are atmospherically distributed and responsive, and these stories are themselves co-composed by the climatic world.

Of course, settlers like myself cannot simply narrate ourselves into this emplaced and intergenerational lifeworld. However, we can be inspired by the power of such collaborative climate stories to cultivate, and continue, non-anthropocentric modes of being. The kinds of cloudy collectives I am storying are somewhat distinct from Bawaka Country's, but hopefully edge us towards modes of being – relational, responsive, caring, more-than-human – that are more closely aligned with Wukun̲ than with the Anthropocene. These cloudy collectives are not particularly unified and they include climate-complicit people.

Emerging from the unsettling atmosphere composed by witnessing each other's ecological distress, our class formed such a cloudy collective. Referring to Australia's high levels of climate denial, early in semester a student commented in class: 'I like Australia, Australia's a cool place. But it's disheartening. You look around, and it's like, where'd everyone go? And they're running away...It's like, [sigh], Jesus guys.' His exasperation at pervasive climate denial led to empathetic laughter from the group. This moment established an in-group based on a common feeling of disappointment, a feeling directed at climate-silent Australians who were thus excluded from the group. This 'affectively contagious, easily shared' and 'we-creating'[29] nature of climate change's affective atmospheres was articulated by another student who identified a 'unanimous feeling of frustration shared by the whole class' at the start of the course. Our class conversations enabled us to share and explore the ways we were encountering climate change, and to witness each other doing so. These collaborative practices that unearthed shared experiences of distress led to students articulating, storying, and thus enacting a moody group into being. One student expressed how connecting over common concerns generated energising sensations: 'I'm so glad I changed into this class – it's more of a climate change therapy group than a university subject.' This followed another student's comment (included in the vignette of Chapter 4) about how national identities of denial require collective practices of climate change therapy, demonstrating how one person's story of climate collectives can spark collective enactments and the consolidation of such modes of being. We did, in some ways, go on to become a kind of climate change therapy group,[30] as the following comment from the anonymous student survey narrates:

> I really valued the ferocious intensity of information that was shared with us. Wednesday became my favourite day because of this subject. A group formed to discuss the challenging aspects of each lecture and this was continued in the tutorial classes. I appreciated that we were dealing with ethically challenging topics, it became a philosophical class about life's most challenging questions, a platform for deeper thinking, ideas and discussion with like-minded people. We had the tutorial and each other to help guide our understandings and relationship with climate change. The class felt like a home for ideas and discussions.

Other students also noted the affective affinities arising from our class practices of encountering, witnessing and storying climate change. One student found that 'being in a classroom full of the people who are going to be having to change the world and just hearing how scared and tired other people are was really cathartic.' For another, being 'surrounded by people who share similar values' and who 'are all about making a change towards a better future' was 'reassuring' and 'empowering.' These comments demonstrate the affective support that connecting with other climate-concerned people can offer. They also speak to the performative capacities of storying climate collectives: to narrate them is to do more than imagine them; it can enact them into being through creating inclusive templates that people inhabit and then reinforce.

However, the group emerging in our class was both inclusive and alienating, at times dissolving and at other times erecting boundaries according to the unfolding affective experiences of different people. Consider one young woman's end-of-semester comment:

> During the course, I felt scared and overwhelmed or emotional and didn't know how to explain the importance of climate change or how to speak about it…From being absolutely motivated and inspired to wanting to shut myself away from the world and everything in between…This class has given me hope, as everyone was so open and happy to share, while I have had a strong feeling of anxiety during this course as I feel everyone is so smart, powerful and brilliant and I'm sometimes too anxious to speak. I think this is a result of how emotional climate change truly is.

This student's comment complicates the too-easy narrative whereby we can enrol people into empowering groups by simply suggesting that they are part of one. This oscillation between inspiration and anxiety, and inclusion and exclusion, exemplifies the uneasiness of cloudy collectives. Storying climate change is a differentiated experience, one that is not equally accessible to all. As this student narrates, other people's confidence flattened her own. Despite the emerging sense of a group, her individual capability to participate in this was overwhelmed by the emotional intensity of climate change coursing through her body, rendering her physically unable to speak, and articulating differences, borders and hierarchies between her and others in the class.

To further complicate the potentially rose-tinted notion of a 'climate change therapy group' spontaneously arising in the course, a student commented in the anonymous end of semester survey that:

> I noticed in a few classes that only the guys were speaking. About five guys would speak in a row and take up quite a bit of 'space.' I think this is a product of a wider societal problem where women's opinions are less valued and women often do not feel confident enough to voice their opinion. A lot of people need a few seconds of silence before they are comfortable to say something.

This story directly articulates the important point that storying climate change emerges with and reconfigures subjectivities and the hierarchies they are entangled with. This student's identification that storying and being witnessed are practices that are distributed along gender binaries, but do not have to be, re-enacts the gender binary in order that it may be undone. It speaks to the ways that gender is made and unmade, reproduced and reworked in partnership with climate change. In these ways, storying climate change contributed to who we were becoming, both as a group and as individuals. As these two students' stories of being alienated due to feeling anxious or uncomfortable demonstrate, the 'whole class' did not share affective experiences at all times. Particular affective regimes sometimes included, and at other times excluded, people from the collective. Cloudy collectives demonstrate that individuality – and thus, hierarchies – can be re-articulated, rather than dissolved, through our enmeshment with others.[31]

In addition to the ambiguous inclusion of students within the class, the cloudy collective sometimes incorporated, and sometimes excluded, people from beyond the classroom cohort. Throughout the semester, students debated the complicated reasons for climate denial and the conceptualisation of denial as an act – rather than an identity – that we engage in to more or less an extent. This served to blur the borders between 'us' and 'them.' In these ways, the boundaries of the group both expanded and contracted, incorporating and expelling people according to affective similarities, differences, empathies and dismissals. But the figure of the cloudy collective explicitly articulates this: whereas the pretence of unity tends to exclude, cloudy collectives are formed by tensions and discordances which can generate metamorphic and transient alliances. These are collectives that 'some people immerse themselves in, or dip in and out of…or build a light and temporary link to before they move on to something else.'[32]

Their moody membranes offer not solidarity so much as ambiguous affinities. Through loosening the requirement for homogenisation and conformity, cloudy collectives can unfurl, extend and disperse beyond the original relations that cultivated them to create proliferating, genuine, but never stable or settled alliances. For example, as the earlier comment about the 'ferocious intensity' of the class noted, a group of students began hanging out together in the one-hour gap between our official class activities, exceeding the bounds of the classroom. Encountering, witnessing and storying climate change can become 'viral response-abilities.'[33] These atmospheric practices can be carried in the dispersing cloudy collective, enacted beyond the original time-place-body entanglement. Another student's end of semester reflection speaks to how cloudy collectives can disseminate templates for others to inhabit:

> This class if anything has spurred my optimism further. We have a group of amazing human beings heading out there setting prime examples of the better people we all can be. The better people we can be, the better our world can be.

Following the conclusion of the course, students and I contributed to this extension, dissipation and reconfiguration of our cloudy collective through the affective-climatic practices we enacted in our distinct communities. Some students took up roles facilitating environmental community building; others wrote and shared poetry; some made documentary films. Another organised a music festival and invited some of us to speak to the punters about climate change; as part of this we made a banner which read 'loving low carbon life' and took it to the People's Climate March in the lead up to the Paris Climate Summit, arriving in multiple but overlapping groups which were unavoidably forced apart amongst the 50,000 strong crowd. Of course, over the longer term we lost contact; and some of the students would never have felt part of this group at all, whether because it was so ethereal as to be imperceptible to them, or because its regimes of affinity were uninviting or even unwelcoming to them.

Further afield, I see cloudy collectives emerging and dissipating all throughout the climate movement. For example, Extinction Rebellion (XR) has choreographed 'cloudy' protests which take their cue from Hong Kong's pro-democracy activists – who had themselves been inspired by Bruce Lee's counsel to 'be water' in order to be 'shapeless,' 'formless' and to be able to 'flow.'[34] XR's adaptation, 'swarms for survival,' involve many small groups of people dispersing throughout a city and then suddenly coalescing to form expansive but short-lived rallies that rapidly dissolve in order to evade police. But XR's signature protest is the 'die-in': playing 'dead fish' in public spaces to symbolise the possibility of human extinction. Activists' frequent experience of burnout arising through participating in such emotionally charged actions has spurred XR to theorise, practice and embed 'regenerative culture' within the movement. Workshops and online resources build rebels' emotional, interpersonal and spiritual capacities in order to ensure longevity in a movement that may otherwise have dwindled rapidly.[35] Beyond XR, during the COVID-19 pandemic, climate activists have organised actions in 'the cloud,' with affect-fuelled online rallies where people can protest alone/together, contributing to this diffuse association. Cloudy collectives operate in less obvious ways too, in places where no unified or organised group can be confidently identified. Through the multitudes of organisations, businesses, schools, councils, communities and collations of individuals filtering through and in-between them, all of us witnessing climate change in myriad ways, we connect, overlap and interpermeate each other, building new relationships around the shared concern climate change generates in us.

***

Climate change is a collective action problem 'par excellence.'[36] The stories people tell of human-climate relations do not just describe or imagine those relations but actively enact them into being. Storying climate collectives contributes to the ways we comprehend and inhabit our transcorporeal bodies. Thus, we need stories that perform empowering templates for collective action but which resist anthropocentrism in all its guises: the erasure of human diversity, the denial of non-human agency, and the fantasy of independence and control. We need stories that enable us to identify as *part of* climate change, and that enable us to stay with the ethical and interpersonal challenges of *living with* it. And we need people to be actively engaged in the composition of such stories, so that they may inhabit, diversify and disperse these ways of relating.

In our climate class, some students said that the practices of collectively and affectively encountering, witnessing and storying climate change contributed to the emergence of a kind of group. Cloudy collectives is my effort to re-story human–climate relationships through my class' collective but differentiated experiences. A cloudy collective is not simply an aggregation of humans but is a more-than-human entanglement where climate change is *the* key participant. Climate change's affective atmospheres contribute to the emergence of these moody, nebulous groups, and these collectives in turn reconfigure atmospheres through their breathy practices of storying climate change. As a changing sense of self, the cloudy collective is evidence of the affective transformation that was unfolding in that class.

Such cloudy collectives do not require coherence or unity. Attuning to the possibilities for collaboration despite and through diversity and differentiation may offer a more empowering, and even hopeful, narrative that people can inhabit. Cloudy collectives might enable people to be part of – but not swallowed by – groups they do not entirely agree or identify with, but which nevertheless share similar concerns. Appreciating that cloudy collectives exist in the relations in-between people, places and planet, and that they are constantly on the move, might also enable us to perceive promising cultural and emotional transformations that are not yet otherwise noticeable.

Through my articulation of cloudy collectives, I hope that you too might be drawn in, and begin to identify the atmospheric alignments between yourself and others – near and far, here and not-yet, human and more-than-human. Of course, cloudy collectives are not the only characters we need. But if we are to learn to live with climate change, we need stories populated by characters that are collective, non-anthropocentric, and that can engage in the affective labour of facing and responding to climate collapse. These characters need to be both relatable and transformative, so that we may inhabit them and change ourselves in the process. Most of all, these stories need to be co-created by those we wish to engage in these transformations, not just told to them. So, how else might we story climate change? What climate care-full characters are you (plural) crafting?

Notes

1 Denise Baden, "Solution-Focused Stories Are More Effective Than Catastrophic Stories in Motivating Proenvironmental Intentions," *Ecopsychology* 11, no. 4 (2019): 254–63, https://doi.org/10.1089/eco.201 9.0023; Abel Gustafson et al., "Personal Stories Can Shift Climate Change Beliefs and Risk Perceptions: The Mediating Role of Emotion," *Communication Reports* 33, no. 3 (2020): 121–35, https://doi.org/10.1080/ 08934215.2020.1799049; Brandi Morris et al., "Stories Vs. Facts: Triggering Emotion and Action-Taking on Climate Change," *Climatic Change* 154, no. 1 (2019): 19–36, https://doi.org/10.1007/s10584-019-02425-6.

2 Marcia McKenzie and Andrew Bieler, *Critical Education and Sociomaterial Practice: Narration, Place, and the Social* (New York, NY: Peter Lang, 2016).

3 Donna Haraway, *Staying with the Trouble: Making Kin in the Chthulucene* (Durham and London: Duke University Press, 2016); Thom van Dooren, Eben Kirksey, and Ursula Münster, "Multispecies Studies: Cultivating Arts of Attentiveness," *Environmental Humanities* 8, no. 1 (2016): 1–23, https://doi.org/10.1215/22011919-3527695.

4 WWF, *Australia's 2019–2020 Bushfires: The Wildlife Toll*, World Wildlife Fund (2020), https://www.wwf.org.au/news/news/2020/3-billion-animals-impacted-by-australia-bushfire-crisis#gs.p9wksk.

5 Kerrin Thomas, Michael Cavanagh, and Kim Honan, "Beekeepers Traumatised and Counselled after Hearing Animals Screaming in Pain after Bushfires," *ABC News*, November 27, 2019. https://www.abc.net.au/

news/2019-11-20/beekeepers-traumatised-by-screaming-animals-after-bushfires/11721756; Naaman Zhou, "Heartbreaking and Heartwarming: Animals Rescued from Australia's Bushfires Devastation," *The Guardian*, December 24, 2019. https://www.theguardian.com/environment/2019/dec/24/heartbreaking-and-heartwarming-animals-rescued-from-australias-bushfires-devastation.

6 Justin Huntsdale, "Bat Clinic Inundated as Wildlife Carers Sign up En Masse Following Bushfire Disaster," *ABC News*, February 27, 2020. https://www.abc.net.au/news/2020-02-27/bat-clinic-inundated-as-wildlife-volunteer-numbers-ride/12003064.

7 Serpil Oppermann, "From Material to Posthuman Ecocriticism: Hybridity, Stories, Natures," in *Handbook of Ecocriticism and Cultural Ecology*, ed. Hubert Zapf (Berlin, Boston: De Gruyter, 2016); James Bradley, "Is It Possible to Write Good Fiction About Climate Change?," (2010), https://cityoftongues.com/2010/03/22/is-it-possible-to-write-good-fiction-about-climate-change/; Timothy Clark, *Ecocriticism on the Edge: The Anthropocene as a Threshold Concept* (London: Bloomsbury, 2015).

8 Mahlu Mertens and Stef Craps, "Contemporary Fiction Vs. The Challenge of Imagining the Timescale of Climate Change," *Studies in the Novel* 50, no. 1 (2018): 136. http://hdl.handle.net/1854/LU-8548225.

9 Ariel Kroon, "Imagining Action In/Against the Anthropocene: Narrative Impasse and the Necessity of Alternatives to Effect Resistance," *The Goose* 1, no. 18 (2020): 2.

10 For a consideration of the ethics of approaching unsettling stories of climate change, see Alison Ravenscroft, "Strange Weather: Indigenous Materialisms, New Materialism, and Colonialism," *The Cambridge Journal of Postcolonial Literary Inquiry* 5, no. 3 (2018), https://doi.org/10.1017/pli.2018.9.

11 Marco Caracciolo and Shannon Lambert, "Narrative Bodies and Nonhuman Transformations," *SubStance* 48, no. 3 (2019); Oppermann, "From Material to Posthuman Ecocriticism: Hybridity, Stories, Natures."

12 Amitav Ghosh, *The Great Derangement: Climate Change and the Unthinkable* (Chicago: The University of Chicago Press, 2016).

13 For example, roughly 70% of Americans surveyed in 2016 said they rarely or never speak about climate change with friends or family. Edward Maibach et al., *Is There a Climate "Spiral of Silence" in America?* (New Haven: Yale University and George Mason University, 2016).

14 Greta Thunberg, *No One Is Too Small to Make a Difference* (Penguin, 2019), 34–5.

15 Andreas Malm and Alf Hornborg, "The Geology of Mankind? A Critique of the Anthropocene Narrative," *The Anthropocene Review* 1, no. 1 (2014).

16 Eileen Crist, "On the Poverty of Our Nomenclature," *Environmental Humanities* 3 (2013).

17 Malm and Hornborg, "The Geology of Mankind? A Critique of the Anthropocene Narrative."

18 Heather Davis and Zoe Todd, "On the Importance of a Date, or Decolonizing the Anthropocene," *ACME: An International E-Journal for Critical Geographies* 16, no. 4 (2017).

19 Haraway, *Staying with the Trouble: Making Kin in the Chthulucene*; Lesley Head, "Contingencies of the Anthropocene: Lessons from the 'Neolithic',"

The Anthropocene Review 1, no. 2 (2014), https://doi.org/10.1177/205301 9614529745.

20 While the notion of the Anthropocene has been widely critiqued, there are far fewer interrogations of the premise of 'human induced,' or 'anthropogenic,' climate change, perhaps because questioning this framing is so often done by climate deniers. While 'human induced' and 'anthropogenic' – rather than 'human caused' – do leave scope for acknowledging that the 'climate is responsive to our nudges only because it is far more precarious than we ever dared imagine,' as Nigel Clark puts it (below), the premise of attributing this 'nudge' to all humans rather than specific industrial systems has all the same problems as the Anthropocene: homogenisation, erasure, assimilation, demoralisation and dehumanisation. Nigel Clark, "Volatile Worlds, Vulnerable Bodies," *Theory, Culture & Society* 27, no. 2–3 (2010): 32, https://doi.org/10.1177/0263276409356000.

21 Haraway, *Staying with the Trouble: Making Kin in the Chthulucene.*

22 Astrida Neimanis, *Bodies of Water: Posthuman Feminist Phenomenology* (London: Bloomsbury Academic, 2017), 40–1.

23 Neimanis, *Bodies of Water: Posthuman Feminist Phenomenology*, 41.

24 The Cloud Appreciation Society, "Manifesto," (2016). https://cloudappreciationsociety.org/manifesto/.

25 Bawaka Country et al., "Gathering of the Clouds: Attending to Indigenous Understandings of Time and Climate through Songspirals," *Geoforum* 108 (2020): 296, 97, https://doi.org/10.1016/j.geoforum.2019.05.017.

26 Bawaka Country et al., "Gathering of the Clouds: Attending to Indigenous Understandings of Time and Climate through Songspirals," 297.

27 Bawaka Country et al., "Gathering of the Clouds: Attending to Indigenous Understandings of Time and Climate through Songspirals," 298.

28 Bawaka Country et al., "Gathering of the Clouds: Attending to Indigenous Understandings of Time and Climate through Songspirals," 300.

29 Ashlee Cunsolo Willox, "Climate Change as the Work of Mourning," *Ethics & the Environment* 17, no. 2 (2012): 145, 49, https://doi.org/10.2979/ethicsenviro.17.2.137.

30 Here, students drew on dominant norms in our society that understand talking about feelings as a practice often conducted with a professional therapist. In line with Maria Ojala's work (below), I reframe this climate change therapy group as a cloudy collective, because we were not engaging in a-political, individualistic therapy so much as politically transgressive collective witnessing that had both cathartic and transformative effects. Nevertheless, this group was not untherapeutic. Maria Ojala, "Facing Anxiety in Climate Change Education: From Therapeutic Practice to Hopeful Transgressive Learning," *Canadian Journal of Environmental Education* 21 (2016).

31 For useful explorations of the intersectional politics of posthuman embodiment, see: Neimanis, *Bodies of Water: Posthuman Feminist Phenomenology*; Magdalena Górska, *Breathing Matters: Feminist Intersectional Politics of Vulnerability* (Linköping University Electronic Press, 2016).

32 Kathleen Stewart, "Atmospheric Attunements," *Environment and Planning D: Society and Space* 29, no. 3 (2011): 452, https://doi.org/10.1068/d9109.

33 David Rousell, Amy Cutter-Mackenzie, and Jasmyne Foster, "Children of an Earth to Come: Speculative Fiction, Geophilosophy and Climate

Change Education Research," *Educational Studies* 53, no. 6 (2017), https://doi.org/10.1080/00131946.2017.1369086.

34 Antony Dapiran, "'Be Water!': Seven Tactics That Are Winning Hong Kong's Democracy Revolution," *New Statesman*, August 1, 2019, https://www.newstatesman.com/world/2019/08/be-water-seven-tactics-are-winning-hong-kongs-democracy-revolution.

35 Christie Wilson, *Regenerative Culture: Regen 101 Workshop Booklet* (Extinction Rebellion, 2020), https://ausrebellion.earth/resources/.

36 Jale Tosun and Jonas Schoenefeld, "Collective Climate Action and Networked Climate Governance," *Wiley Interdisciplinary Reviews: Climate Change* 8, no. 1 (2017): 2, e440, https://doi.org/10.1002/wcc.440.

References

Baden, Denise. "Solution-Focused Stories Are More Effective Than Catastrophic Stories in Motivating Proenvironmental Intentions." *Ecopsychology* 11, no. 4 (2019): 254–63. https://doi.org/10.1089/eco.2019.0023.

Bawaka Country, S. Wright, S. Suchet-Pearson, K. Lloyd, L. Burarrwanga, R. Ganambarr, M. Ganambarr-Stubbs, B. Ganambarr, and D. Maymuru. "Gathering of the Clouds: Attending to Indigenous Understandings of Time and Climate through Songspirals." *Geoforum* 108 (2020): 295–304. https://doi.org/10.1016/j.geoforum.2019.05.017.

Bradley, James. "Is It Possible to Write Good Fiction About Climate Change?". (2010). https://cityoftongues.com/2010/03/22/is-it-possible-to-write-good-fiction-about-climate-change/.

Caracciolo, Marco, and Shannon Lambert. "Narrative Bodies and Nonhuman Transformations." *SubStance* 48, no. 3 (2019): 45–63.

Clark, Nigel. "Volatile Worlds, Vulnerable Bodies." *Theory, Culture & Society* 27, no. 2–3 (2010): 31–53. https://doi.org/10.1177/0263276409356000.

Clark, Timothy. *Ecocriticism on the Edge: The Anthropocene as a Threshold Concept*. London: Bloomsbury, 2015.

Crist, Eileen. "On the Poverty of Our Nomenclature." *Environmental Humanities* 3 (2013): 129–47.

Cunsolo Willox, Ashlee. "Climate Change as the Work of Mourning." *Ethics & the Environment* 17, no. 2 (2012): 137–64. https://doi.org/10.2979/ethicsenviro.17.2.137.

Dapiran, Antony. "'Be Water!': Seven Tactics That Are Winning Hong Kong's Democracy Revolution." *New Statesman*, August 1, 2019. https://www.newstatesman.com/world/2019/08/be-water-seven-tactics-are-winning-hong-kongs-democracy-revolution.

Davis, Heather, and Zoe Todd. "On the Importance of a Date, or Decolonizing the Anthropocene." *ACME: An International E-Journal for Critical Geographies* 16, no. 4 (2017): 761–80.

Ghosh, Amitav. *The Great Derangement: Climate Change and the Unthinkable*. Chicago: The University of Chicago Press, 2016.

Górska, Magdalena. *Breathing Matters: Feminist Intersectional Politics of Vulnerability*. Linköping University Electronic Press, 2016.

Gustafson, Abel, Matthew Ballew, Matthew Goldberg, Matthew Cutler, Seth Rosenthal, and Anthony Leiserowitz. "Personal Stories Can Shift Climate Change Beliefs and Risk Perceptions: The Mediating Role of Emotion." *Communication Reports* 33, no. 3 (2020): 121–35. https://doi.org/10.1080/08934215.2020.1799049.

Haraway, Donna. *Staying with the Trouble: Making Kin in the Chthulucene*. Durham and London: Duke University Press, 2016.

Head, Lesley. "Contingencies of the Anthropocene: Lessons from the 'Neolithic'." *The Anthropocene Review* 1, no. 2 (2014): 113–25. https://doi.org/10.1177/2053019614529745.

Huntsdale, Justin. "Bat Clinic Inundated as Wildlife Carers Sign up En Masse Following Bushfire Disaster." *ABC News*, February 27, 2020. https://www.abc.net.au/news/2020-02-27/bat-clinic-inundated-as-wildlife-volunteer-numbers-ride/12003064.

Kroon, Ariel. "Imagining Action in/against the Anthropocene: Narrative Impasse and the Necessity of Alternatives to Effect Resistance." *The Goose* 1, no. 18 (2020): 1–12.

Maibach, Edward, Anthony Leiserowitz, Seth Rosenthal, Connie Roser-Renouf, and Matthew Cutler. *Is There a Climate "Spiral of Silence" in America?* New Haven: Yale University and George Mason University, 2016.

Malm, Andreas, and Alf Hornborg. "The Geology of Mankind? A Critique of the Anthropocene Narrative." *The Anthropocene Review* 1, no. 1 (2014): 62–9.

McKenzie, Marcia, and Andrew Bieler. *Critical Education and Sociomaterial Practice: Narration, Place, and the Social*. New York, Bern, Berlin, Bruxelles, Frankfurt am Main, Oxford, Wien: Peter Lang, 2016.

Mertens, Mahlu, and Stef Craps. "Contemporary Fiction Vs. The Challenge of Imagining the Timescale of Climate Change." *Studies in the Novel* 50, no. 1 (2018): 134–53. http://hdl.handle.net/1854/LU-8548225.

Morris, Brandi, Polymeros Chrysochou, Jacob Dalgaard Christensen, Jacob Orquin, Jorge Barraza, Paul Zak, and Panagiotis Mitkidis. "Stories Vs. Facts: Triggering Emotion and Action-Taking on Climate Change." *Climatic Change* 154, no. 1 (2019): 19–36. https://doi.org/10.1007/s10584-019-02425-6.

Neimanis, Astrida. *Bodies of Water: Posthuman Feminist Phenomenology*. London: Bloomsbury Academic, 2017.

Ojala, Maria. "Facing Anxiety in Climate Change Education: From Therapeutic Practice to Hopeful Transgressive Learning." *Canadian Journal of Environmental Education* 21 (2016): 41–56.

Oppermann, Serpil. "From Material to Posthuman Ecocriticism: Hybridity, Stories, Natures." In *Handbook of Ecocriticism and Cultural Ecology*, edited by Hubert Zapf. Berlin, Boston: De Gruyter, 2016.

Ravenscroft, Alison. "Strange Weather: Indigenous Materialisms, New Materialism, and Colonialism." *The Cambridge Journal of Postcolonial Literary Inquiry* 5, no. 3 (2018): 353–70. https://doi.org/10.1017/pli.2018.9.

Rousell, David, Amy Cutter-Mackenzie, and Jasmyne Foster. "Children of an Earth to Come: Speculative Fiction, Geophilosophy and Climate Change Education Research." *Educational Studies* 53, no. 6 (2017): 654–69. https://doi.org/10.1080/00131946.2017.1369086.

Stewart, Kathleen. "Atmospheric Attunements." *Environment and Planning D: Society and Space* 29, no. 3 (2011): 445–53. https://doi.org/10.1068/d9109.

The Cloud Appreciation Society. "Manifesto." (2016). https://cloudappreciationsociety.org/manifesto/.

Thomas, Kerrin, Michael Cavanagh, and Kim Honan. "Beekeepers Traumatised and Counselled after Hearing Animals Screaming in Pain after Bushfires." *ABC News*, November 27, 2019. https://www.abc.net.au/news/2019-11-20/beekeepers-traumatised-by-screaming-animals-after-bushfires/11721756.

Thunberg, Greta. *No One Is Too Small to Make a Difference.* Penguin, 2019.

Tosun, Jale, and Jonas Schoenefeld. "Collective Climate Action and Networked Climate Governance." *Wiley Interdisciplinary Reviews: Climate Change* 8, no. 1 (2017): 1–17 e440. https://doi.org/10.1002/wcc.440.

van Dooren, Thom, Eben Kirksey, and Ursula Münster. "Multispecies Studies: Cultivating Arts of Attentiveness." *Environmental Humanities* 8, no. 1 (2016): 1–23. https://doi.org/10.1215/22011919-3527695.

Wilson, Christie. *Regenerative Culture: Regen 101 Workshop Booklet.* Extinction Rebellion, 2020.

WWF. *Australia's 2019–2020 Bushfires: The Wildlife Toll.* World Wildlife Fund (2020). https://www.wwf.org.au/news/news/2020/3-billion-animals-impacted-by-australia-bushfire-crisis#gs.p9wksk.

Zhou, Naaman. "Heartbreaking and Heartwarming: Animals Rescued from Australia's Bushfires Devastation." *The Guardian*, December 24, 2019. https://www.theguardian.com/environment/2019/dec/24/heartbreaking-and-heartwarming-animals-rescued-from-australias-bushfires-devastation.

6 Conclusion: Bearing worlds

Aiming to articulate an affective and relational pedagogy for climate change engagement, across the last three chapters I have explored the practices of encountering, witnessing and storying climate change. These three practices are quite distinct from *knowing about* climate change. They are practices that we enact *with and as part of* climate change. We do this through our emplaced-and-dispersed bodies. This conceptualization of human knowledge and agency (as well as embodiment and subjectivity) as constantly emerging with climate change is central to the ontological dimensions of learning to live with climate change: appreciating that climate is living-with, and that we are entangled with/in this patterned set of relations.

I have focused on how these practices can enable us to engage with and respond to the distress of climate change. Encountering climate anxiety, witnessing multiple climate realities, and storying climate collectives are key methods through which we can engage in the affective dimensions of learning to live with climate change. Through a detailed exploration of one university class, I have demonstrated how collectively and reiteratively engaging in these practices can move us from anxiety towards affective transformation. In that class, encountering, witnessing and storying each other encountering, witnessing and storying climate change enabled us to identify, share, normalise, explore and respond to our ecological distress. Throughout the semester, climate change's affective agency decomposed our anthropocentric individualities, and in their place a promising, if elusive, climate capable collective was emerging. We engaged in alternative ways of relating to each other and to climate change, and through doing so, formed a 'cloudy collective': a moody assemblage of climate-changed humans. As Anna Tsing describes, we were 'forced to be ever more aware of the process of finding allies and building collaborations when we realize[d] we are not the crest of a wave to an

DOI: 10.4324/9780367441265-6

imagined better future.'[1] We were learning – implicitly and explicitly, consciously and subconsciously – that who we are and how we live is always affected by and entangled with climate, and we were also increasing our collective capacities to engage with the unsettling realities of climate collapse.

If we are to catalyse the extensive changes required for a viable climate future, we must engage both the affective and ontological dimensions of learning to live with climate change. Beginning from an understanding that *climate is patterns of affect that we are entangled with* clarifies that climate anxiety emerges from our transcorporeal enmeshment with disordered planetary atmospheres. It also emphasises that this disorder is itself generated by particular affective regimes: the anthropocentric infatuation with humans, the colonial zeal for extractivism, the neoliberal obsession with individual wealth, and the patriarchal disdain for care. Attuning to the enmeshment of human emotions and global climate in this way demonstrates the need to interrogate and reconfigure the emotional norms and expectations of climate-complicit cultures. This is what is required if we are to work towards emotional climate justice.

In contrast to this approach, experiences of climate anxiety are often framed as internal, psychological human phenomena and emotional resilience is advocated as the appropriate antidote. The problem with this is that focusing on emotional resilience risks reaffirming and re-centring the needs, identities and feelings of climate-complicit people and could enable us to 'bounce back' to anthropocentric ways of being. To do this would be to reinstate the affective regimes that cause the problem. For these reasons, I advocate understanding climate anxiety not (solely) as a problem to be addressed through emotional resilience but as an opportunity for, and even part of the process of, affective transformation.

Affective transformation encompasses, differs from and exceeds emotional resilience. Rather than fortifying a pre-existing self against ecological distress, affective transformation is about changing form through engagement with unravelling climatic relations. Affective transformation involves a subjective metamorphosis, a changing of the sense of self from an insulated individual human being to a distributed, atmospheric, more-than-human 'becoming.' It is about attuning to our climatic transcorporeality, in the sense of being both porously enmeshed with climate change and dynamically changing because of this. We need to consider ourselves as atmospherically dispersed and climatically entangled beings who are composed, decomposed and recomposed by the patterns of the planet's energetic cycles. We are always becoming part of the climate because 'the ebb and flow of meteorological life transits through us, just as the actions, matters, and meanings of our own bodies return to the climate in myriad ways.'[2] Equally and relatedly, we are always changing through our relations with climate: we are becoming-with climate.[3] Affective transformation is therefore a process through which we become (with) climate change, transforming our sense of self and our understandings of where our 'selves' begin, end, emerge, extend and dissolve. It is a means through which we 'lose our former selves,' but it also has a 'we-creating' potential.[4]

Relatedly, affective transformation involves an ability to prepare for and endure climate change: to get on, to make do, to manage and cope, through changing our expectations and organising ourselves in ways that cultivate more collective mettle. It takes a nod from trans-formative climate adaptation which argues that responding to the impacts of climate change should compel the reorganisation of the social systems that created the problem: we need to bounce elsewhere, not bounce back.[5] Focusing specifically on the relational, emotional, and embodied ways that we are and could be differently entangled with the more-than-human world, affective transformation speaks to our capacity to cool planetary climates through encountering, navigating and trying to bear the affronts of global heating. It is an ability to endure ongoing interpersonal reconfiguration, an openness to and capacity to abide emotional challenges, a reworking of our affective expectations, skills, repertoires, routines and relations. Affective transformation recognises that future ways of living-with will be radically different to those we have come to know and/or love, and involves grieving for the losses we are already experiencing and those that are yet to come. This is not about becoming resigned to climate change, giving up or thinking that it is too late to do anything.

Affective transformation contributes to processes of 'making worlds at the end of the world,'[6] or what I term *bearing worlds*. Bearing worlds involves *enduring* the pain that current and potential climate change engenders, while *labouring* to generate desirable and possible, though always uncertain and indeterminate, futures. This is what makes affective transformation more hopeful than emotional resilience: affec-tive transformation requires that climate-complicit peoples change themselves and the socio-economic structures they are entangled with, for it is only through such processes that we might create more promising worlds. For those of us who are both vulnerable to and complicit with climate change, learning to live with climate change 'cannot be about fortifying our own havens,' rather, it 'requires interrupting our existing patterns of weathermaking.'[7] Somewhere between the bleak acceptance of resilience and the anthropocentric delusions of saving the planet is a space of humble resolve which can enable us to keep facing up to the terrors of climate change so that we might prevent the worst. Learning to live with climate change thus refers to how

> the making of worlds and the sense of the end of a certain kind of world coincide: Here are new articulations of subjects, relations and environments that are going on and unfolding, not always with a plan, but still settling into particular lifelines that inform the possibilities of other worlds to come.[8]

If we are to engage in these labours of bearing worlds, we have to learn to live with climate change. This involves learning with climate change that life arises through relations. Living is always living-with. There is no existence outside of ecology, and our human experiences and identities are intimately enmeshed with the 'environment.' Learning to live with climate change is a deep attunement to the entanglement of all life and the cultivation of appropriate ways of relating to and engaging with that world. Of course, in an era of rapid planetary breakdown, any effort to acknowledge interconnection will identify that valuable relations are threatened, and thus involve grief for those losses.[9] Therefore, learning to live with climate change also involves the kinds of respectful sorrow encompassed in the common idiom 'learning to live with' something. It refers to the entwined affective labours of identifying and mourning relationships as they are torn apart, disfigured and/or reconfigured as the planet cooks.

As such, learning to live with climate change is going to be disconcerting and distressing, but it could also be joyful, reassuring, refreshing and/or invigorating. If engaged with carefully, together, in respectful communities, it can be regenerative. Learning to live with climate change is a metamorphosis which will no doubt be tricky, painful and probably ugly and confusing at times, but that could lead to necessary and beneficial changes so as to better relate, integrate and work with the changing world in ways that enable multispecies survival. In these ways, learning to live with climate change also involves continuing to act for a future which is desirable despite being different, or perhaps acting for a future that is less bad than it would have been if we did not act. It is an informed practice that yearns for and creates more livable climate futures. By engaging with the varied ways that we are entangled with climate change, we can develop more embodied, animated, ecological and responsible ways of inhabiting and enacting our shared world.

But learning to live with climate change is not simple, straightforward, able to be rolled out through standardised programs or ticked off as though it could be completed or mastered. There are no clearly definable steps or measurable goals. This is partly because the climate has already changed, and so there are no 'solutions' to climate change, just better and worse (non)responses which become available and dissipate according to particular relationalities that are constantly unfolding and reconfiguring. Climate change has been described as a 'super wicked problem'[10] and the 'greatest moral challenge of our time,'[11] but I am not sure any words we have available can do justice to the political, social, personal, emotional, ethical, intellectual and ecological complexities and challenges facing us, in all our different ways. Learning to live with climate change recognises the myriad, overlapping, compounding and continuously morphing situations that climate change poses, which are unfair, painful and unresolvable, but which still demand our best efforts. While limiting global warming to 1.5 degrees might still be achievable and is certainly more desirable than 2 or 3 or 4, 5 or 6 degrees,[12] the 1 degree we have already experienced is horrifying, and our best-case scenarios still involve massive losses, unconscionable injustices, complex and traumatic compromises, and a whole lot of unpredictable, uncontrollable and unmanageable eventualities. We cannot 'reverse' climate change; even if we can reduce global average temperatures, the world's complex suite of relations will be irrevocably ruptured.[13] Learning to live with climate change acknowledges that we will be living with climate change in some ways or others, no matter how coordinated or ethical our collective actions may be; and that we – in all our myriad forms of 'we' – have to find ways to keep going despite this. We will have to continue grappling with the ever-shifting complexities of climate change, doing our (entangled) best to reduce emissions and adapt to the impacts, all while our collapsing worlds disrupt our sense of self.

Ultimately, learning to live with climate change is an existential (i.e., affective and ontological) task of composing not just different lifestyles, but different conceptions of what life is, what it means to live, and how to live well. This will no doubt be a 'bewildering' process.[14] We cannot completely predict what is necessary, what will eventuate, who and what will affect us, or how we – individually and collectively – will become (with) climate change. We do not, and cannot, fully know where we are going; it is about learning, and I am learning, too.

***

So then, what are we to do? Yes, this is overwhelming, yes, it is an emerging process, but we have to start somewhere, and we need people to take action, right? Of course. However, the three practices that I have discussed in this book – encountering, storying and witnessing climate change – are actions. They require active participation; it takes energy, time, and embodied work to engage in them. It takes emotional labour to come face to face with the multiple realities of climate change, to grapple with them and to speak candidly about them. It takes effort and courage to show up to classes – or public workshops, protests, family dinners or workplace meetings – and acknowledge the climate crisis and stand up for serious action. And it is physically exhausting to do such things on an ongoing basis. This is the work of mourning that Ashlee Cunsolo argues is constitutive of climate change.[15] Attuning to the embodied energy these tasks demand can help us better support people to begin and continue engaging with climate change.

Further, encountering, witnessing and storying climate change are more than actions. Or, to be more precise, they operate differently to how we normally understand climate action and they provide pause to carefully reconsider this. Efforts to cultivate climate action typically presume that the actor is an individual human who is capable of exercising power over the world to achieve a predetermined expert-prescribed aim. Encountering, witnessing and storying are not actions in this sense. Rather, they are what Karen Barad terms *intra-actions*: like interactions in that they are a process of mutually intermeshing with and effecting each other, but more 'intra' than 'inter' because we are already, and always, part of climate.[16] Encountering, witnessing and storying climate change are practices through which climate change acts on us as much as, and while, we act on it, and where we collectively act-with each other. Through enacting these practices, climate change affects us, and we become climate-changed. Which is to say, they make us different. In this sense, they are also climate *diffractions*: different actions, actions that emerge from difference, actions that differentiate, actions that generate difference.[17]

This is the kind of action we need. We urgently need action, but specifically, we need actions that emerge from and contribute to appreciation of our intimate enmeshment with, and vulnerability to, climate change, which is to say, actions that challenge and reorient cultures of dominance. So long as we believe that accurately informed and sufficiently concerned individual humans can successfully manage the climate, then even if we do stabilise average planetary temperatures we are destined to simply displace the problems of anthropocentrism elsewhere. We need actions that rapidly reduce greenhouse gas emissions but which do not allow us to engage in the fallacy that we can comprehensively predict, model and manage the planet's temperature with precision as though it were an oven. While we might start with intentions and directions, the world will bump into, disrupt, reinforce, invert and/or displace our desired trajectories and inclinations, perhaps blocking or resisting our efforts, but also potentially generating and/or amplifying novel, different, unanticipated forms of climate action.

That, of course, is unsettling. Letting go of control sounds like giving up. It is not. Letting go of the myth of climate control in no way requires that we back down in fights against institutionalised climate denial or systematic destruction of planetary life. In fact, decentring the individual human and attuning to collective, more-than-human agencies in many ways multiplies or expands our capacities, even as it might temper our intentional power. Recalling that climate is a patterned set of relations and climate change a disruption in those patterns infers that responding to climate change requires reconfiguring those disturbing relations.[18] Thus, changing our relationships is one of the most systemic ways we can change extractive cultures. Cultivating climate responsibility is therefore about increasing the capacity of 'individuals' to enter into, and attend to (to become aware of, to nurture and to respond to), relationships with human and more-than-human others. But changing relationships requires being in relationship, which means working in partnership, not being in control, and therefore not being able to pre-empt where such collaborations might go.[19] By working in relation with our more-than-human worlds, we can cultivate and leverage coalitions that honour the agency of non-humans and re-situate ourselves as part of the climate, contributing to humble and ecologically responsive cultures.

The kinds of (intra-active and diffractive) actions that are needed to prevent and adapt to climate collapse are unlimited. Encountering, witnessing and storying climate change are practices that can catalyse and inform such responses. Learning to live with climate change will occur through various modes of encountering, witnessing and storying climate change: our existing ways of life will be countered, as we collectively but differentially witness climate change unfolding. We will tell stories of these climate-human-world reconfigurations, and we will need new stories to inhabit, if we are to learn to live with climate change. Because these practices spark and shape other climate actions, it matters how climate change is encountered, it matters whether and in what ways climate change is witnessed and it matters which climate stories are told.

***

Because these practices are so emotionally demanding, there is an important role for people to encourage, support and guide themselves and others to engage in them. This final section explores how encountering, witnessing and storying climate change can be facilitated. But before I discuss tangible strategies, it is important to flag that anyone who tries to engage others in these practices is not separate from those people and their entanglements with climate change. All such facilitators already are, and will become further, enmeshed with those they seek to engage. Facilitators, whether they be teachers, therapists, activists, community members or otherwise, will be drawn into encountering, witnessing and storying climate change themselves. They will be subject to the emotional intensities of climate change and will also have their individual autonomy countered through this process. This is because we can only ever act *with* and *as part of*, not *on*, climate change, and climate change is not just 'out there' but threaded through our everyday worlds. Even when we try to engage 'others' in the processes of learning to live with climate change, our interpersonal relationships, and thus our identities and sense of self, will continue to be challenged and reconfigured. Indeed, climate change is enrolling us in these practices all the time; the task for facilitators is to help people consciously attune to, dwell within, inhabit and amplify them. Without this careful attention, these practices may be so disorienting that they stimulate avoidance or hyper-anthropocentric responses that seek to gain control of the world.

I find the work of engaging others in the processes of learning to live with climate change immensely exhilarating and rewarding. But it is also incredibly challenging. It often leaves me feeling confused, exhausted, demoralised and useless. Other environmental educators have similar experiences. Consider this educator's description of their attempt to support their students to engage with their climate anxiety:

> I gave the space in the class for students to express what they feel and their anger. But it came to a dead-end, with everyone's hope ending very low. I could feel their frustration and hopelessness, but struggled finding a way to channel it into anything

constructive without sounding naive by saying positive messages that don't mean much given the challenges we face.[20]

Referring to a similar experience, another educator that my co-researchers and I surveyed mentioned that after their class one day they 'had a long cry on my commute home, and wound up cancelling plans I had to meet friends that evening.' Such experiences demonstrate that trying to support others to engage with and navigate their own ecological distress often leads to feelings of inadequacy and despair becoming contagious.

The conceptual frameworks offered in this book, specifically of atmospheres as both climatic and affective, help make sense of this. The notion of 'space' – and often, 'safe space' – is commonly used by facilitators to name the conditions necessary for people to feel capable of engaging with issues that are potentially controversial or traumatic (as demonstrated by the educator quoted earlier). I suggest *unsettling and regenerative atmospheres* are a better way of conceptualising the conditions necessary for engaging people in the labours of bearing worlds. The conditions required do not begin and end with certain spatial features such as classroom walls, nor can they be generated without the active participation of the people involved. Rather, they emerge from the continuously unfolding relations people have with each other, places and the planet which extend in and spiral through vast temporal and geographical scales. Atmospheres, in this sense, are the prevailing affective regimes which co-emerge with the people and climates participating in them. Focusing on atmospheres, rather than spaces, emphasises that the emotional intensities that climate change generates in some of us emanate outwards, envelop, interpermeate and are re-made by others. Through their capacity to connect us with human and more-than-human others these atmospheres reconfigure us, and through our reassembled relations, we participate in the ongoing composition of atmospheres, and therefore, climates. Atmospheres also emphasise that the affective transformation unfolding with/in such relationships can leach out and extend beyond the initial spacetime, for example, in the form of a cloudy collective.

While we need such atmospheres to be safe enough that people feel supported to engage with the phenomenal violences of climate change, they cannot be so safe that they insulate and excuse us from our complicity in these violences. Rather than provide safety or therapy that soothes, we need to support and activate people so that they can 'stay with the trouble' of climate change, as Donna Haraway puts it.[21] Etymologically, to trouble means to stir up, disturb and make cloudy.[22] Unsettling atmospheres can 'stir people up' such that we can shake loose of the complacency of the status quo through collectively expressing our anger, grief and guilt. We need to dwell with the unjust histories, uncomfortable presents, and uncontrollable futures that we are unavoidably enmeshed with, so as to inspire alternative modes of being. Doing this is immensely challenging, and so we need to ensure the atmospheres we compose are also regenerative. Regenerative atmospheres do not allow moral or existential comfort other than that dependent on radical cultural change, but they provide the conditions for people to contemplate, experiment with, rehearse and enact that change. Unsettling and regenerative atmospheres therefore create an affective refuge, not from climate change but from climate denial, and this can enable us to collectively recuperate and recompose ourselves, and in turn, the climate.

Unsettling and regenerative atmospheres can be enacted through collectively encountering, witnessing and storying climate change, where climate change is understood to be *right here*, in our interpersonal relations and everyday lives, as well as distributed throughout massive geographical and temporal scales. This requires providing opportunities to encounter, witness and story grand disruptions, as well as to encounter, witness and story *each other* as we do so. Unsettling and regenerative atmospheres are composed when people are supported to engage with and respond to multiple experiences of climate change overlapping, intersecting, compounding and diffracting each other. In order to (co-)compose such atmospheres, designating specific times and spaces to encounter, witness and story climate change can be effective. However, such transformative atmospheres can also be (co-)composed spontaneously in everyday situations. If a designated time is available, facilitators can offer stimulus to focus people's attention, such as a video about climate change, but this is not necessary. Participants have often already encountered, witnessed and storied climate change in their lives and these existing experiences can be attuned to, and subsequently encountered, witnessed and storied – collectively, if appropriate. The key is asking questions, or designing activities, that encourage and support people to consider, dwell with, explore and respond to their own, and others', affective entanglements with climate change. Ensuring these atmospheres are both unsettling and regenerative requires a critical and compassionate approach from facilitators. This is a delicate balance made all the harder by the fact that facilitators are not the only ones contributing to atmospheres. Further, atmospheres are perceived and experienced differently by different people, meaning what might feel unsettling for one person can be deeply traumatic for another. Checking in with people and providing resources for further support are always necessary.[23]

In my efforts to support people to learn to live with climate change, I never start by trying to convince people about climate change, make them concerned about it or check they understand the science. Instead, I work with people's existing relations and experiences and begin by asking people how climate change, as they understand it, makes them feel. There are many ways this question can be explored and responded to, whether in a one-on-one conversation, group discussion, through drawing, writing or other kinds of artistic practice. Sometimes it is too intense to approach this directly and accompanying it with an incidental activity that can diffuse the focus, such as making handicrafts or going for a walk, can be beneficial. Whatever the approach, the point is not to judge or try to change how people feel, because there is no correct way to feel about climate change. The point is just to listen: to witness them as they encounter, witness and story their own experiences. Bearing witness – and it is a *labour* – and validating their feelings, whatever they are, demands empathy and compassion and it builds trust and respect. From there, collectively, you (plural) can explore the relations – beneficial or destructive, enduring or disappearing – that contribute to those feelings, and explore ways of responding that might address the pain in both the short and long term, such as connecting them to particular activist and/or support networks. But rushing to response is not always wise; sometimes we have to sit with feelings over time before the appropriate pathway becomes apparent.

Responding to others in distress is really hard. It is difficult because there are no easy solutions to offer and the inability to alleviate another's pain can be demoralising and debilitating; adding the requirement for affective unsettlement into the mix makes for incredibly complex work. However, whatever responses we can – and *will* – fashion will be spurred by our collective grief, anxiety, anger, guilt, love and hope. Uncovering, exploring, dwelling with and channelling these feelings is the foundation for any effort, small or grand, to begin and continue responding to climate change. Therefore, supporting people to engage with the anguish of climate change is crucial, 'as uncomfortable as it is, and as untrained we might feel to manage it.'[24] As Susanne Moser writes, learning to be with people in distress is a crucial skill for environmental leadership,[25] and this includes learning to live with our own distress. As unsettling as this work is, it is through this relational labour that we will birth more liveable worlds.

Collectively exploring the affective intensities of climate change, in all their uncanny, disconcerting complexity, and experimenting with new ways of identifying with each other and climate, can help us reinvent ourselves and regenerate our worlds. These emotional, relational, more-than-human practices collectively contribute to the ontological, sociological and ecological tasks of learning to live with climate change. This is an open-ended and unpredictable process which is going to be traumatic, both because of the disasters wrought by climate change, and also because disidentifying from deep-seated cultural norms is painful. Yet, through collectively engaging in these unsettling processes, we just might learn, with climate change, how to live together with our temperamental planet.

Notes

1 Anna Lowenhaupt Tsing, "Getting by in Terrifying Times," *Dialogues in Human Geography* 8, no. 1 (2018): 75, https://doi.org/10.1177/20438206 17738836.

2 Astrida Neimanis and Rachel Loewen Walker, "Weathering: Climate Change and the 'Thick Time' of Transcorporeality," *Hypatia* 29, no. 3 (2014): 560, https://doi.org/10.1111/hypa.12064.

3 Donna Haraway, *When Species Meet* (Minneapolis: University of Minnesota Press, 2008).

4 Ashlee Cunsolo Willox, "Climate Change as the Work of Mourning," *Ethics & the Environment* 17, no. 2 (2012): 145, 49, https://doi.org/10.2979/ethicsenviro.17.2.137.

5 Libby Porter et al., "Climate Justice in a Climate Changed World," *Planning Theory & Practice* 21, no. 2 (2020), https://doi.org/10.1080/1464 9357.2020.1748959.

6 Jennifer Gabrys, "Making Worlds at the End of the World," *Dialogues in Human Geography* 8, no. 1 (2018): 63, https://doi.org/10.1177/20438206177 38830.

7 Astrida Neimanis and Jennifer Mae Hamilton, "Weathering," *Feminist Review* 118, no. 1 (2018): 82, https://doi.org/10.1057/s41305-018-0097-8.

8 Gabrys, "Making Worlds at the End of the World," 61.

9 Cunsolo Willox, "Climate Change as the Work of Mourning."

10 Richard Lazarus, "Super Wicked Problems and Climate Change: Restraining the Present to Liberate the Future," *Cornell Law Review* 94 (2008).

11 Marc Hudson, "It's Ten Years since Rudd's 'Great Moral Challenge', and We Have Failed It," *The Conversation*, March 31, 2017. https://theconversation.com/its-ten-years-since-rudds-great-moral-challenge-and-we-have-failed-it-75534.

12 IPCC, *Global Warming of 1.5°C: Summary for Policy Makers*, Intergovernmental Panel on Climate Change (2018), http://www.ipcc.ch/report/sr15/.

13 Average global surface temperature is just one indicator of climate change. Climate change, as I understand it, refers to the disruption in complex patterns of multispecies-and-inorganic relations, which, once ruptured, cannot be reassembled. For a discussion on related challenges of reducing climate change to a singular measure, see Mike Hulme, "Climate Emergency Politics Is Dangerous," *Issues in Science and Technology* 36, no. 1 (2019).

14 Nathan Snaza, "Bewildering Education," *Journal of Curriculum and Pedagogy* 10, no. 1 (2013), https://doi.org/10.1080/15505170.2013.783889.

15 Cunsolo Willox, "Climate Change as the Work of Mourning."

16 Karen Barad, *Meeting the Universe Halfway: Quantum Physics and the Entanglement of Matter and Meaning* (Durham and London: Duke University Press, 2007).

17 Karen Barad, "Diffracting Diffraction: Cutting Together-Apart," *Parallax* 20, no. 3 (2014), https://doi.org/10.1080/13534645.2014.927623; Donna Haraway, *Modest−Witness@Second−Millennium.Femaleman−Meets−Oncomouse: Feminism and Technoscience* (New York and London: Routledge, 1997); Blanche Verlie and CCR15, "From Action to Intra-Action? Agency, Identity and 'Goals' in a Relational Approach to Climate Change Education," *Environmental Education Research* 26, no. 9–10 (2020), https://doi.org/10.1080/13504622.2018.1497147.

18 Kyle Whyte, "Too Late for Indigenous Climate Justice: Ecological and Relational Tipping Points," *WIREs Climate Change* 11, no. 1 (2020), https://doi.org/10.1002/wcc.603.

19 Verlie and CCR15, "From Action to Intra-Action? Agency, Identity and 'Goals' in a Relational Approach to Climate Change Education."

20 For the whole study, see Blanche Verlie et al., "Educators' Experiences and Strategies for Responding to Ecological Distress," *Australian Journal of Environmental Education* (2020), https://doi.org/10.1017/aee.2020.34.

21 Donna Haraway, *Staying with the Trouble: Making Kin in the Chthulucene* (Durham and London: Duke University Press, 2016).

22 Haraway, *Staying with the Trouble: Making Kin in the Chthulucene,* 1.

23 Useful books in this regard include: Sarah Jaquette Ray, *A Field Guide to Climate Anxiety: How to Keep Your Cool on a Warming Planet* (Oakland: University of California Press, 2020); Anouchka Grose, *A Guide to Eco-Anxiety: How to Protect the Planet and Your Mental Health* (London: Watkins Media, 2020). For some people, professional support will be necessary.

24 Sarah Jaquette Ray, "Coming of Age at the End of the World: The Affective Arc of Undergraduate Environmental Studies Curricula," in *Affective Ecocriticism: Emotion, Embodiment, Environment*, ed. Kyle Bladow and Jennifer Ladino (University of Nebraska Press, 2018), 301.

25 Susanne Moser, "Getting Real About It: Meeting the Psychological and Social Demands of a World in Distress," in *Environmental Leadership*, ed.

Deborah Gallagher (Los Angeles, London, New Delhi, Singapore, Washington, DC: SAGE Publications, 2012).

References

Barad, Karen. "Diffracting Diffraction: Cutting Together-Apart." *Parallax* 20, no. 3 (2014): 168–87. https://doi.org/10.1080/13534645.2014.927623.

Barad, Karen. *Meeting the Universe Halfway: Quantum Physics and the Entanglement of Matter and Meaning.* Durham and London: Duke University Press, 2007.

Cunsolo Willox, Ashlee. "Climate Change as the Work of Mourning." *Ethics & the Environment* 17, no. 2 (2012): 137–64. https://doi.org/10.2979/ethicsenviro.17.2.137.

Gabrys, Jennifer. "Making Worlds at the End of the World." *Dialogues in Human Geography* 8, no. 1 (2018): 61–3. https://doi.org/10.1177/2043820617738830.

Grose, Anouchka. *A Guide to Eco-Anxiety: How to Protect the Planet and Your Mental Health.* London: Watkins Media, 2020.

Haraway, Donna. *Modest−Witness@Second−Millennium.Femaleman−Meets−Oncomouse: Feminism and Technoscience.* New York and London: Routledge, 1997.

Haraway, Donna. *Staying with the Trouble: Making Kin in the Chthulucene.* Durham and London: Duke University Press, 2016.

Haraway, Donna. *When Species Meet.* Minneapolis: University of Minnesota Press, 2008.

Hudson, Marc. "It's Ten Years since Rudd's 'Great Moral Challenge', and We Have Failed It." *The Conversation*, March 31, 2017. Accessed October 16, 2018. https://theconversation.com/its-ten-years-since-rudds-great-moral-challenge-and-we-have-failed-it-75534.

Hulme, Mike. "Climate Emergency Politics Is Dangerous." *Issues in Science and Technology* 36, no. 1 (2019): 23–5.

IPCC. *Global Warming of 1.5°C: Summary for Policy Makers.* Intergovernmental Panel on Climate Change (2018). http://www.ipcc.ch/report/sr15/.

Lazarus, Richard. "Super Wicked Problems and Climate Change: Restraining the Present to Liberate the Future." *Cornell Law Review* 94 (2008): 1153–234.

Moser, Susanne "Getting Real About It: Meeting the Psychological and Social Demands of a World in Distress." In *Environmental Leadership*, edited by Deborah Gallagher, 900–8. Los Angeles, London, New Delhi, Singapore, Washington, DC: SAGE Publications, 2012.

Neimanis, Astrida, and Jennifer Mae Hamilton. "Weathering." *Feminist Review* 118, no. 1 (2018): 80–4. https://doi.org/10.1057/s41305-018-0097-8.

Neimanis, Astrida, and Rachel Loewen Walker. "Weathering: Climate

Change and the 'Thick Time' of Transcorporeality." *Hypatia* 29, no. 3 (2014): 558–75. https://doi.org/10.1111/hypa.12064.

Porter, Libby, Lauren Rickards, Blanche Verlie, Karyn Bosomworth, Susie Moloney, Bronwyn Lay, Ben Latham, Isabelle Anguelovski, and David Pellow. "Climate Justice in a Climate Changed World." *Planning Theory & Practice* 21, no. 2 (2020): 293–321. https://doi.org/10.1080/14649357.2020.1 748959.

Ray, Sarah Jaquette. *A Field Guide to Climate Anxiety: How to Keep Your Cool on a Warming Planet.* Oakland: University of California Press, 2020.

Ray, Sarah Jaquette. "Coming of Age at the End of the World: The Affective Arc of Undergraduate Environmental Studies Curricula." In *Affective Ecocriticism: Emotion, Embodiment, Environment*, edited by Kyle Bladow and Jennifer Ladino, 219–399. University of Nebraska Press, 2018.

Snaza, Nathan. "Bewildering Education." *Journal of Curriculum and Pedagogy* 10, no. 1 (2013): 38–54. https://doi.org/10.1080/15505170.2013.783889.

Tsing, Anna Lowenhaupt. "Getting by in Terrifying Times." *Dialogues in Human Geography* 8, no. 1 (2018): 73–6. https://doi.org/10.1177/204382061 7738836.

Verlie, Blanche, and CCR15. "From Action to Intra-Action? Agency, Identity and 'Goals' in a Relational Approach to Climate Change Education." *Environmental Education Research* 26, no. 9–10 (2020): 1266–80. https://doi.org/10.1080/13504622.2018.1497147.

Verlie, Blanche, Emily Clark, Tamara Jarrett, and Emma Supriyono. "Educators' Experiences and Strategies for Responding to Ecological Distress." *Australian Journal of Environmental Education* Online First (2020): 1–15. https://doi.org/10.1017/aee.2020.34.

Whyte, Kyle. "Too Late for Indigenous Climate Justice: Ecological and Relational Tipping Points." *WIREs Climate Change* 11, no. 1 (2020): e603. https://doi.org/10.1002/wcc.603.

Appendix: Discussion questions

The following questions may be useful to work through to process, translate, activate or respond to this book. You might like to respond in creative ways, through art, dance, poetry, place-based meditation and/or enact them as modes of political engagement, for example through engaging in these discussions in public places, forming your responses into public art installations or letters to be sent to political representatives. Or, you might just want to sit with them yourself. They can also be used as examples of the kinds of prompts you could use when facilitating climate change engagement for others.

Don't forget to reach out to others, including professionals, if you or someone you know needs that kind of support.

How does climate change make you feel? Which of these feelings can you name, and can you identify the broader socio-political regimes that contribute to you feeling this way?

Are there other sensations that you cannot name, that are more fleeting, or too uncanny, that climate change generates in you?

What relations do these sensations bring your awareness to – what do they emerge from, and what do they contribute to?

What relations, futures and/or opportunities do your feelings indicate are (becoming) closed or lost to you because of climate change? Can these losses be channelled into openings – not necessarily solutions, or restoration, but promising possibilities?

What might be adequate or at least somewhat constructive ways of responding to these feelings? If you don't know, where could you go, who could you connect with, what could you do, in order to develop some strategies?

Are there other ways that you are encountering climate change? Which of your relations are being disrupted, and how is this challenging and reconfiguring you?

What are the ways that you are witnessing climate change? What kinds of knowledge do they generate, and what relations do they enrol you in?

Are there modes of witnessing climate change that you are not engaging with? Why? What might happen if you did engage in these?

What stories do you tell yourself, and others, about climate change? How do you tell them – through conversation, body language, your actions or something else?

What worlds – ways of identifying, knowing, relating and interacting – are these stories summoning into being? Are they the kinds of world you want to bring forth? What other stories might you tell?

In what ways are you becoming (with) climate change – how are you changing through your entanglement with, and as part of, planetary heating and ecosystem collapse? Are there promising dimensions to this – even as challenging as they might be?

Relatedly, how are your actions, your body, your desires, your relationships, becoming part of the climate? What might all this have to say about how we understand ourselves (as individuals and as humans)? Are there any benefits that might arise from this?

How would you describe the atmospheres – as circulating and dispersing patterns of energy that have both emotional and climatic dimensions or implications – that you are participating in or bumping up against? Who and what is energised by them, and who is smothered? How could you intervene in or reshape them?

Can you identify emotional, embodied, and/or affective dimensions in other people's relations with and participation in climate change? How might thinking through your bodily and affective experience help empathise with those who are and are not like you?

What affinities might be established around this – what would you need to do in order to curate some kind of collective climate response?

How might you encourage and support your communities to engage in the processes of learning to live with climate change? Have you asked them how they are feeling? Have you let them know how you are feeling?

If, or when, you have shared how you are feeling: what did you learn, and what happened, through and after doing so?

Index

Page numbers with an "n" refer to the note.

action/acting 5, 9, 24, 29–31, 90,
94–5, 103–4, 114–18;
see also agency; diffraction; intra-
action; protest; strike
adaptation 113, 115, 117
affect: defined 24; differentiated
experiences of 28–9, 101–2; as
energies and intensities 6; as not
fully knowable 57–8; as planetary
embodied entanglement 1, 28–9,
32, 49, 57–8, 77, 120; relation with/
distinction from emotions 56;
relation with knowledge 27–8,
71–3, 77–8, 80; and temporality 33,
57, 77 see also atmospheres;
affective regimes of extractivism;
affective transformation; agency
affective regimes of extractivism 7, 9,
24, 56, 112
affective transformation 8–10, 12, 24,
93, 104, 111–13, 119
agency: of climate 4–5; climate's
affective 11, 23–4, 28, 51, 54, 59,
72, 79, 93, 98, 111; climate's
spiritual 78; defined 4; emerging
through human–climate relations
111; of humans 4, 91, 95–96;
humans unleashing climate's 26; of
non–humans 104, 117; personal 58;
of stories 91
Ahmed, Sara 29, 58
animals 22, 27, 31, 68–9, 91

Anthropocene 95–6, 98–9
anthropocentrism: and accounts of
reality 68, 70, 79; defined 4, 104;
distress as opportunity to rethink
55, 58; posthumanism as challenge
to 15n15; and saving the planet
113, 116; and stories 93, 96, 105
anxiety 49–52; as barrier to action 2,
8; contribution to collective
subjectivity 96–7, 101; as entangled
with regimes of affect 7, 112;
guidance on how to manage 17n45;
as opportunity for transformation
and action 112, 121; relation with
and capacity to disrupt
anthropocentrism, modernity,
individualism 9–10, 51–5, 58–9;
relation with smoke 22, 32; as
relational and transcorporeal
phenomenon 49, 51–2, 112;
relationship with knowledge 49, 73;
of scientists 72; teachers' efforts to
address 118
apathy 2, 8–9, 70, 74, 88–9;
see also denial
aspiration 4, 12, 28;
see also breath; hope
atmospheres: as affective and climatic
5, 9, 22–5, 48, 57–8, 97, 99, 104, 119;
in classrooms 48; humans entangled
with and part of 6–7, 9, 12, 24–5, 99,
112; as metaphor for vibe 2, 23; as

part of and indicator of climate system 3; productive of collectives 97, 99, 102, 104; of protests 29; settler and racialised 28–9; as spatially diffuse and composed by and productive of bodies 28–32, 48–9, 57–8, 104, 112, 119; unsettling and/or regenerative 10, 99, 119–20
authorship 11, 92–3
autonomy 5, 11, 26, 98, 118

Bawaka Country research collective 4, 98–99
becoming 6, 12, 15n15, 50, 59, 73, 77, 93, 96–101, 112–16
binary *see* dualism
bodies: and affect 24, 57; and anxiety 49–50, 52, 55; climate and/as bodies, their relations and experience 5–7, 9, 24–8, 32, 48, 52, 58, 111, 113; differentiated experiences of climate change 5, 28–9; encountering, witnessing and/or storying as embodied practice 9, 58, 73, 104, 111, 116; imaginaries 54; and knowledge systems e.g. science 2–4, 28, 69–72, 77–8, 99; and ways of relating 113–14; *see also* transcorporeality
breath, or inhibition of it 12, 22, 25–30, 33, 37n49, 52, 75, 78, 95, 97–8, 104; *see also* inspiration; aspiration
business as usual 12, 74–7; *see also* capitalism; extractivism; neoliberal
burnout 50, 103; *see also* exhaustion

capitalism 7, 53, 66, 75, 89 *see also* extractivism
care 2, 112; *see also* self-care
children 29–30, 75–8; *see also* young people; strike
climate: defined 5–7; as flows of affect/energy 12, 24, 30, 48, 98, 112
climate change: as disruption in patterns of relation 7, 54, 117; as distinct from global warming 115, 122n13

climate-control; *see* control
clouds 1, 31; to make cloudy 119; *see also* cloudy collectives
cloudy collectives 11, 30–1, 93, 96–104, 111, 119
collective: climate change as a collective action problem 47, 58, 92, 104; collective engagement with feelings 10, 13, 71, 96, 99–101, 104, 111, 118–22; collectives emerging from distress 29, 31, 55–6, 58, 73, 80, 111, 119–21; need for collective stories of action 92–6, 107n30; *see also* cloudy collectives
colonial, colonialism *see* extractivism; neoliberal; settler; decolonial
complicity 7–9, 96–9, 112–13, 119
control 49; as anthropocentrism 4–5, 52–3, 118; climate change disrupting human 7, 22, 26–7, 52, 55–9, 78–9, 115; 'climate-control' 48–9, 58; letting go of as ethical practice 59, 67, 93, 104, 117
culture *see* extractivism; media
cultural burning 32
Cunsolo, Ashlee 6, 55, 116

decolonial, decolonisation 16n22, 31, 53; *see also* unsettle
denial: as an act not an identity 102; climate science and/as 67–72; contribution to experience of double realities 74–7; as coping mechanism 2, 8, 65; of embodied knowledge 4; entangled with identity 47, 66, 100; mass, institutionalised and systemic 1–2, 9, 29, 32, 55, 99–100, 117; of non-human agency 104; refuge from 120
denier 29, 46, 55, 66, 74
despair *see* hopelessness
diffraction 75, 116–17, 120
disasters 5, 32–33, 50, 77
disembodied *see* bodies
disempowerment 2, 54; *see also* hopelessness
disidentification 58, 122
distress 48–51, 121

diversity 96, 104; *see also* bodies
domination 32, 49, 53;
    see also control
dreams 51–52; daydreams and
    nightmares 27, 75–77;
    see also unreal
dualism 4, 73, 101

embodied *see* bodies
emergency 30, 32; *see also* disasters
emissions 4, 26, 33, 97, 115;
    see also energy
emotional resilience 8, 12, 112–13
empathy 8, 48, 72, 121
emplaced 28, 48, 67, 73, 77–8,
    98–9, 111
encountering 46–59; capacities of 96,
    99–100, 102, 104, 111, 113; defined
    49–50; as diffractive and intra-
    active actions 116–17; as embodied
    practice enacted with climate
    change 9, 26, 33, 111; facilitators'
    role 118–21; as methodology 10–11
end of worlds: 12, 31, 50–53, 58,
    113–15
energy: energised people 11, 57, 90,
    94, 100; planetary and embodied
    enmeshment with and dispersal of
    6, 28, 32, 48, 52, 97, 112; systems
    33; *see also* affect; climate; labour
entanglement: cloudy collectives as
    96, 102–4; of facilitators and those
    they work with 118; of humans and
    climate 5–6, 8, 24–5, 28–9, 48, 55,
    58, 73, 77–8, 111–13; storying as
    practice of 101; with other people
    11, 28, 31
ethics 52, 67–9, 73, 88, 100, 104, 115
exhaustion 2, 11, 29, 90, 116
existential 2, 28, 31, 47, 50–1, 58, 72,
    115, 120
experience: of climate as emerging
    from social, planetary and
    transcorporeal relations 2, 5, 24,
    28, 48–9, 52, 77; of climate change
    as inexplicable 57–9; of climate
    scientists 71; of environmental
    educators 118–19; lived experience
    as nuanced and unique 67, 79; as

mechanism for alternate climate
    relations 72, 77–8; need to witness
    others' 72, 77; shared but
    differentiated 28–9, 31, 101, 104,
    120; stories of as relatable and
    limits to this 91–2; of weather
    versus climate 70, 81n13;
    see also reality; unreal
extinction (mass, human) 46, 91,
    95, 103
Extinction Rebellion 67, 103
extractivism 4, 7, 9, 32, 59, 68–9, 72,
    75, 79, 96, 112, 117

feeling: climate as 24; as connective
    and alienating 98–9, 101;
    entanglement of climate and 25, 98
fiction 90–2
fire 1, 9, 22–33, 50, 56–7, 67–9, 75–8,
    91, 98
frustration 8, 53–6, 65, 89, 94,
    98–9, 118
futures: entanglement with past and
    present 77–8, 90, 98–9, 119; and
    progress narratives 52–3, 58;
    promising 3, 94, 100, 112–14;
    stolen 52–3; visions of 88–9; worry
    about 27, 32, 48, 51, 54–5, 65, 74

gaslighting 29, 77, 80
gender 67–8, 101
God's eye view 3, 69, 70, 72–3, 78–9;
    see also vision
grief 2, 27–32, 48–9, 52–6
guilt 50–6

Haraway, Donna 69, 72, 81n19, 119
Head, Lesley 48, 52
hierarchy 28, 32, 73, 101–2
hope 53, 73, 93–5, 98, 113, 118;
    see also energy; aspiration;
    inspiration; regeneration
hopelessness 54, 96, 98
Hulme, Mike 3–4, 122n13
human-climate relations 1–3, 6, 8–9,
    24, 79, 93, 104, 117;
    see also humanity; climate
humanity, human nature 7, 53, 65,
    88–90, 95–6, 98

identification 11, 58, 74, 96, 98, 101, 104, 121; *see also* identity
identity 47, 50, 66, 74, 90; *see also* sense of self; subjectivity; becoming
inclusive 94–5, 97, 100–102
indigenous 4–5, 7, 16n22, 29, 31, 52–3, 78–9, 89–90, 95, 98–9; *see also* colonial; decolonial; settler
individual: and emotions and affect 24; erosion of sense of individuality 10–11, 50–7, 98, 101, 102, 111–12, 118; individualism 3–4, 8–9, 12, 68, 89, 94–6, 112, 116–17; perspectives 91–2
inequality 28, 73
inspiration 3, 5, 9, 10, 48, 88, 90, 95, 99, 101, 120; *see also* hope; breath
intra-action 116–17
intergenerational 7, 29, 78–9, 98–9

justice 5–6, 9–10, 29, 32, 50, 58, 68, 73, 79, 98, 112, 115

knowledge 2–5, 9, 26–8, 31, 49, 51, 56, 58, 65, 67–80, 99, 111, 115

labour 14, 29, 68–72, 80, 96, 105, 114, 116
learning to live with climate change 5, 8, 12–4, 111–22
living-with 5–8, 104, 111–115
loss 48, 52–5, 113–15

Matrix, the 74
media 22–33, 56, 68, 92
mental illness/health 8, 50
metamorphosis 58, 97, 112, 114 *see also* affective transformation; becoming
misanthropy 96
more-than-human: climate as 6; human as 5–7, 11, 112, 114, 117

Neimanis, Astrida 5, 83n35, 107n31
neoliberal 11, 53, 58, 95, 112; *see also* extractivism
nightmares *see* dreams
Norgaard, Kari 8, 74

objectification 6, 49, 73, 78–9
objectivity 69–70
omnicide 7
ontology 79, 91, 111–12, 115, 121
overwhelm 8, 53–7

pedagogy 2–3, 10, 49, 111
planetary: disruption 31, 53, 72, 91, 95; entanglement 5, 7, 28, 52, 57, 58, 90, 97, 112
politics 29–32, 50–5, 71–2, 77, 107n30
posthumanism 4–5, 15n15; *see also* more-than-human
power 28, 32, 49, 92, 116, 117 *see also* politics
prosthesis 28, 73
protest 12–13, 29–30, 103, 116; *see also* strike

reality 32, 65–80, 91, 94–5, 111–12, 116
refuge 120
regeneration 10, 68, 80, 91, 114, 119–121
relationality 5–6
resilience *see* emotional resilience
responsibility 16n36, 55, 93, 96, 98, 117

safe space 119
scepticism *see* denier; denial
science 2–4, 26–8, 32–3, 51, 65, 67, 69–72, 78–9, 120; scientists 54, 75–6
self *see* sense of self
self-care 12
sense of self 8–11, 25, 29, 32, 50–8, 78, 92, 104, 112–15, 118; *see also* becoming; identity; identification; subjectivity
settler 5, 7, 16n22, 25, 28–9, 31, 52, 58, 78–9, 99; *see also* extractivism; unsettle
smoke 1, 22–33, 57, 75, 98
social constructionism 3–4, 9
space, spatial 48, 53, 57–8, 66, 101, 103, 118–20
spirit 4, 78–9, 103
storying 88–105; defined 90–91; as

diffractive and intra-active 116–17; as embodied practice enacted with climate change 9, 33, 111; facilitators' role 118–21; as methodology 10–11
strike 1, 13, 30, 52, 56, 58
subjectivity 11, 25, 91, 96, 101, 111–12; *see also* sense of self; identity

technology 26, 69, 78
testimony 27, 68, 72
therapy 66, 100–1, 107n30, 118–19
Thunberg, Greta 25, 52, 67, 94
transcorporeality: climate's affect as 48–52, 57–8, 77, 80, 98, 112; defined 26; knowledge through 71–3; *see also* bodies
transformation *see* affective transformation
truth 2, 67–8, 74–5, 79
Tuana, Nancy 5, 72–3

uncertainty 51, 58, 65, 113
unity 94–7, 99, 102–4
unreal, unrealities 74–9

unsettle 7–8, 10–11, 47, 49, 52–3, 65, 71, 79, 112, 117, 121–2; unsettling atmospheres 99, 119–20

violence 7–8, 29, 53, 72–3, 98, 119
vision 6, 70–5, 78–9, 81n19, 89–90, 95
vulnerability 7–8, 16n36, 28, 65, 89, 96, 98, 113, 116

weather: animated 78–9; experience 1–2, 29, 50, 76; and relation to climate 3, 70, 81n13; weathering 5, 28, 97, 113
witnessing 65–80; capacities of 96, 99–100, 102, 104; defined 68–9; as diffractive and intra-active actions 116–17; as embodied practice enacted with climate change 9, 33, 111; facilitators' role 118–21; as methodology 10–11
Wright, Alexis 5, 78

young people 13, 29–30, 47, 52, 90, 94; *see also* children; Thunberg, Greta